Energy, Environment, and Sustainability

This books series publishes cutting edge monographs and professional books focused on all aspects of energy and environmental sustainability, especially as it relates to energy concerns. The Series is published in partnership with the International Society for Energy, Environment, and Sustainability. The books in these series are editor or authored by top researchers and professional across the globe. The series aims at publishing state-of-the-art research and development in areas including, but not limited to:

- Renewable Energy
- Alternative Fuels
- Engines and Locomotives
- Combustion and Propulsion
- Fossil Fuels
- Carbon Capture
- Control and Automation for Energy
- Environmental Pollution
- Waste Management
- Transportation Sustainability

More information about this series at http://www.springer.com/series/15901

Avinash Kumar Agarwal
Anirudh Gautam · Nikhil Sharma
Akhilendra Pratap Singh
Editors

Methanol and the Alternate Fuel Economy

Springer

Editors
Avinash Kumar Agarwal
Department of Mechanical Engineering
Indian Institute of Technology Kanpur
Kanpur, Uttar Pradesh, India

Nikhil Sharma
Department of Mechanical Engineering
Indian Institute of Technology Kanpur
Kanpur, Uttar Pradesh, India

Anirudh Gautam
Research Design and Standards Organisation
Ministry of Railways
Lucknow, Uttar Pradesh, India

Akhilendra Pratap Singh
Department of Mechanical Engineering
University of Wisconsin-Madison
Madison, WI, USA

ISSN 2522-8366 ISSN 2522-8374 (electronic)
Energy, Environment, and Sustainability
ISBN 978-981-13-3286-9 ISBN 978-981-13-3287-6 (eBook)
https://doi.org/10.1007/978-981-13-3287-6

Library of Congress Control Number: 2018961230

This Springer imprint is published by the registered company Springer Nature Singapore Pte Ltd.
The registered company address is: 152 Beach Road, #21-01/04 Gateway East, Singapore 189721, Singapore

Preface

Energy demand has been rising remarkably due to the increasing population and urbanization. Global economy and society are significantly dependent on the energy availability because it touches every facet of human life and its activities. Transportation and power generation are two major examples. Without the transportation by millions of personalized and mass transport vehicles and availability of 24 × 7 power, human civilization would not have reached contemporary living standards.

The International Society for Energy, Environment and Sustainability (ISEES) was founded at Indian Institute of Technology Kanpur (IIT Kanpur), India, in January 2014 with the aim of spreading knowledge/awareness and catalysing research activities in the fields of energy, environment, sustainability and combustion. The society's goal is to contribute to the development of clean, affordable and secure energy resources and a sustainable environment for the society and to spread knowledge in the above-mentioned areas and create awareness about the environmental challenges, which the world is facing today. The unique way adopted by the society was to break the conventional silos of specializations (engineering, science, environment, agriculture, biotechnology, materials, fuels, etc.) to tackle the problems related to energy, environment and sustainability in a holistic manner. This is quite evident by the participation of experts from all fields to resolve these issues. ISEES is involved in various activities such as conducting workshops, seminars and conferences in the domains of its interest. The society also recognizes the outstanding works done by the young scientists and engineers for their contributions in these fields by conferring them awards under various categories.

The second international conference on "Sustainable Energy and Environmental Challenges" (SEEC-2018) was organized under the auspices of ISEES from 31 December 2017 to 3 January 2018 at J N Tata Auditorium, Indian Institute of Science Bangalore. This conference provided a platform for discussions between eminent scientists and engineers from various countries including India, USA, South Korea, Norway, Finland, Malaysia, Austria, Saudi Arabia and Australia. In this conference, eminent speakers from all over the world presented their views related to different aspects of energy, combustion, emissions and alternative energy resources for sustainable development and cleaner environment. The conference presented five high-voltage plenary talks from globally renowned experts on topical themes, namely "Is It Really the End of Combustion Engines and Petroleum?" by Prof. Gautam Kalghatgi, Saudi Aramco; "Energy Sustainability in India: Challenges and Opportunities" by Prof. Baldev Raj, NIAS Bangalore; "Methanol Economy: An Option for Sustainable Energy and Environmental Challenges" by Dr. Vijay Kumar Saraswat, Hon. Member (S&T), NITI Aayog, Government of India; "Supercritical Carbon Dioxide Brayton Cycle for Power Generation" by Prof. Pradip Dutta, IISc Bangalore; and "Role of Nuclear Fusion for Environmental Sustainability of Energy in Future" by Prof. J. S. Rao, Altair Engineering.

The conference included 27 technical sessions on topics related to energy and environmental sustainability including 5 plenary talks, 40 keynote talks and 18 invited talks from prominent scientists, in addition to 142 contributed talks, and 74 poster presentations by students and researchers. The technical sessions in the conference included Advances in IC Engines: SI Engines, Solar Energy: Storage, Fundamentals of Combustion, Environmental Protection and Sustainability, Environmental Biotechnology, Coal and Biomass Combustion/Gasification, Air Pollution and Control, Biomass to Fuels/Chemicals: Clean Fuels, Advances in IC Engines: CI Engines, Solar Energy: Performance, Biomass to Fuels/Chemicals: Production, Advances in IC Engines: Fuels, Energy Sustainability, Environmental Biotechnology, Atomization and Sprays, Combustion/Gas Turbines/Fluid Flow/Sprays, Biomass to Fuels/Chemicals, Advances in IC Engines: New Concepts, Energy Sustainability, Waste to Wealth, Conventional and Alternate Fuels, Solar Energy, Wastewater Remediation and Air Pollution. One of the highlights of the conference was the rapid-fire poster sessions in (i) Energy Engineering, (ii) Environment and Sustainability and (iii) Biotechnology, where more than 75 students participated with great enthusiasm and won many prizes in a fiercely competitive environment. More than 200 participants and speakers attended this four-day conference, which also hosted Dr. Vijay Kumar Saraswat, Hon. Member (S&T), NITI Aayog, Government of India, as the chief guest for the book release ceremony, where 16 ISEES books published by Springer, under a special dedicated series "Energy, Environment, and Sustainability" were released. This is the first time that such significant and high-quality outcome has been achieved by any society in India. The conference concluded with a panel discussion on "Challenges, Opportunities & Directions for Future Transportation Systems", where the panellists were Prof. Gautam Kalghatgi, Saudi Aramco; Dr. Ravi Prashanth, Caterpillar Inc.; Dr. Shankar Venugopal, Mahindra and Mahindra; Dr. Bharat Bhargava, DG, ONGC Energy Center; and Dr. Umamaheshwar, GE Transportation, Bangalore. The panel discussion was moderated by Prof. Ashok Pandey, Chairman, ISEES. This conference laid out the road map for technology development, opportunities and challenges in energy, environment and sustainability domains. All these topics are very relevant for the country and the world in the present context. We acknowledge the support received from various funding agencies and organizations for the successful conduct of the second ISEES conference SEEC-2018, where these books germinated. We would therefore like to acknowledge SERB, Government of India (special thanks to Dr. Rajeev Sharma, Secretary); ONGC Energy Center (special thanks to Dr. Bharat Bhargava); TAFE (special thanks to Sh. Anadrao Patil); Caterpillar (special thanks to Dr. Ravi Prashanth); Progress Rail, TSI, India (special thanks to Dr. Deepak Sharma); Tesscorn, India (special thanks to Sh. Satyanarayana); GAIL, Volvo; and our publishing partner Springer (special thanks to Swati Meherishi).

The editors would like to express their sincere gratitude to a large number of authors from all over the world for submitting their high-quality work in a timely manner and revising it appropriately at short notice. We would like to express our special thanks to Dr. Atul Dhar, Dr. Jai Gopal Gupta and Dr. Pravesh Chandra Shukla, who reviewed various chapters of this monograph and provided their valuable suggestions to improve the manuscripts.

At this stage of technology development, transportation and power generation systems are dependent on conventional fuels such as mineral diesel and gasoline, which resulted in the rapid depletion of petroleum reserves. The application of different alternative fuels such as biofuels, alcohols and other synthetic fuels needs to be explored for sustainable development of the automotive sector. Amongst these fuels, the use of methanol has gained significant attention for transportation sector. Therefore, this monograph included several chapters for methanol utilization in IC engine application. This monograph is intended for practitioners working in the energy sector, and we hope that the book would be of great interest to the professionals and postgraduate students involved in fuels, IC engines and environmental research. The main objective of this monograph is to present the status of energy sector, potential alternative fuels and the technologies for promoting the utilization of these alternative fuels.

Kanpur, India — Avinash Kumar Agarwal
Lucknow, India — Anirudh Gautam
Kanpur, India — Nikhil Sharma
Madison, USA — Akhilendra Pratap Singh

Contents

Part I General

1 **Introduction of Methanol and Alternate Fuel Economy** 3
Avinash Kumar Agarwal, Anirudh Gautam, Nikhil Sharma and Akhilendra Pratap Singh

Part II Methanol Economy

2 **Methanol as an Alternative Fuel for Diesel Engines** 9
Hardikk Valera and Avinash Kumar Agarwal

3 **Improving Efficiency of Diesel Traction and Adoption of Liquid Sunshine for Indian Railways** 35
Anirudh Gautam, Vagish Kumar Mishra and Avinash Kumar Agarwal

4 **Enabling Rural Economy in India to Partially Substitute Petroleum Products by Methanol—Technology Solutions and Policy Support** 65
M. S. Srinivasan

Part III Alternative Fuels

5 **Study of Performance and Emissions of Engines Fueled by Biofuels and Its Blends** 77
Gaurav Dwivedi, Suyesh Pillai and Anoop Kumar Shukla

6 **Sustainability Assessment of Biodiesel Production in India from Different Edible Oil Crops Using Emergy Analysis** 107
Shyamal Das, Rahul Dev Misra and Biplab Das

7 **Impact of Tri-Fuel on Compression Ignition Engine Emissions: Blends of Waste Frying Oil–Alcohol–Diesel** 135
Thokchom Subhaschandra Singh and Tikendra Nath Verma

8 Review on the Use of Essential Oils in Compression Ignition Engines 157
S. M. Ashrafur Rahman, T. J. Rainey, Z. D. Ristovski, A. Dowell, M. A. Islam, M. N. Nabi and R. J. Brown

Part IV Utilization Aspects

9 Laser-Ignited Engine Development for Adaptation to Hydrogen-Enriched Compressed Natural Gas (HCNG) 185
Rajesh Kumar Prasad and Avinash Kumar Agarwal

10 Particulate Matter and Its Impact on Human Health in Urban Settings 213
Dev Prakash Satsangi and Avinash Kumar Agarwal

Editors and Contributors

About the Editors

Avinash Kumar Agarwal is Professor in the Department of Mechanical Engineering in Indian Institute of Technology Kanpur. His areas of interest are IC engines, combustion, alternative fuels, conventional fuels, optical diagnostics, laser ignition, HCCI, emission and particulate control, and large bore engines. He has published 24 books and more than 230+ international journal and conference papers. He is fellow of SAE (2012), ASME (2013), ISEES (2015) and INAE (2015). He received several awards such as prestigious Shanti Swarup Bhatnagar Award-2016 in engineering sciences; Rajib Goyal Prize-2015; NASI-Reliance Industries Platinum Jubilee Award-2012; INAE Silver Jubilee Young Engineer Award-2012; SAE International's Ralph R. Teetor Educational Award-2008; INSA Young Scientist Award-2007; UICT Young Scientist Award-2007; INAE Young Engineer Award-2005.

Anirudh Gautam is Executive Director of the Research Designs & Standards Organization, Ministry of Railways, in Lucknow, India. He has completed his master's in engine systems from the University of Wisconsin-Madison (USA) and his Ph.D. from IIT Kanpur, India. After working on the maintenance and operations of locomotives in the Indian Railways, he moved to the manufacture of diesel locomotives at Diesel Locomotive Works in Varanasi, India, where he worked on the diesel engines for locomotives. He was instrumental in developing the first indigenous EMD design locomotive in India and has also developed the ALCO locomotive electronic fuel injection system and the mobile emission test car. His main areas of interest are energy production devices, fuel cells, hybrid power trains and sustainable motive power systems, control systems development and structures optimization. He has been the recipient of many awards and is now working on developing locomotive engines for alternative fuel sources and increased fuel and emission efficiency.

Nikhil Sharma is CSIR-Pool Scientist in the Department of Mechanical Engineering, IIT Kanpur. He has completed his M.Tech. and Ph.D. from NIT Hamirpur (India) and IIT Kanpur, respectively. His research interests include optical diagnostics, fuel spray characterization, emission measurement and management and application of alternative fuels for internal combustion engines.

Akhilendra Pratap Singh is Indo-US Postdoctoral Fellow in the University of Wisconsin-Madison, USA. He received his M.Tech. and Ph.D. in mechanical engineering from Indian Institute of Technology Kanpur, India, in 2010 and 2017, respectively. He worked as CSIR-Pool Scientist at ERL, IIT Kanpur, from 2014 to 2017. His areas of research include advanced low-temperature combustion, optical diagnostics with special reference to engine endoscopy and PIV, combustion diagnostics and engine emissions measurement. He has edited 5 books and published 17 chapters and over 30 peer-reviewed international journal and conference papers.

Contributors

Avinash Kumar Agarwal Engine Research Laboratory, Department of Mechanical Engineering, Indian Institute of Technology Kanpur, Kanpur, India

S. M. Ashrafur Rahman Biofuel Engine Research Facility, Queensland University of Technology (QUT), Brisbane, Australia

R. J. Brown Biofuel Engine Research Facility, Queensland University of Technology (QUT), Brisbane, Australia

Biplab Das Department of Mechanical Engineering, National Institute of Technology Silchar, Silchar, Assam, India

Shyamal Das Department of Mechanical Engineering, National Institute of Technology Silchar, Silchar, Assam, India

A. Dowell Southern Cross Plant Science, Southern Cross University, Lismore, NSW, Australia

Gaurav Dwivedi School of Mechanical Engineering, VIT University, Vellore, India

Anirudh Gautam Research Designs & Standards Organisation, Ministry of Railways, Lucknow, India

M. A. Islam Biofuel Engine Research Facility, Queensland University of Technology (QUT), Brisbane, Australia

Vagish Kumar Mishra Research Designs & Standards Organisation, Ministry of Railways, Lucknow, India

Rahul Dev Misra Department of Mechanical Engineering, National Institute of Technology Silchar, Silchar, Assam, India

M. N. Nabi Central Queensland University, Perth, WA, Australia

Suyesh Pillai Amity University Uttar Pradesh, Noida, India

Rajesh Kumar Prasad Engine Research Laboratory, Department of Mechanical Engineering, Indian Institute of Technology Kanpur, Kanpur, India

T. J. Rainey Biofuel Engine Research Facility, Queensland University of Technology (QUT), Brisbane, Australia

Z. D. Ristovski Biofuel Engine Research Facility, International Laboratory for Air Quality and Health, Queensland University of Technology (QUT), Brisbane, Australia

Dev Prakash Satsangi Indian Institute of Technology Kanpur, Kanpur, India

Nikhil Sharma Indian Institute of Technology Kanpur, Kanpur, India

Anoop Kumar Shukla Amity University Uttar Pradesh, Noida, India

Akhilendra Pratap Singh University of Wisconsin-Madison, Madison, USA

M. S. Srinivasan Indian Institute of Technology, Madras, Chennai, India

Thokchom Subhaschandra Singh Department of Mechanical Engineering, National Institute of Technology Manipur, Langol, Imphal, Manipur, India

Hardikk Valera Engine Research Laboratory, Department of Mechanical Engineering, Indian Institute of Technology Kanpur, Kanpur, India

Tikendra Nath Verma Department of Mechanical Engineering, National Institute of Technology Manipur, Langol, Imphal, Manipur, India

Part I
General

Chapter 1
Introduction of Methanol and Alternate Fuel Economy

Avinash Kumar Agarwal, Anirudh Gautam, Nikhil Sharma and Akhilendra Pratap Singh

Abstract Currently, more than 80% of global energy is supplied through fossil fuels, in which more than 95% of fossil fuel energy is utilized in the transport sector. This has resulted in two issues, namely rapid depletion of petroleum reserves and environmental pollution due to excessive consumption of these petroleum-based fuels. Therefore, it becomes necessary to explore alternative fuels such as methanol, biofuels, compressed natural gas (CNG), hydrogen and other synthetic fuels. These alternative fuels can be used in all transportation modes including roadways as well as railways. This monograph describes different aspects related to these alternative fuels, especially methanol, which has emerged as a potential alternative fuel for both compression ignition (CI) and spark ignition (SI) engines. Utilization of methanol in large-bore engines, use of laser ignition in engines fuelled with gaseous alternative fuels and particulate emission characteristics of engines fuelled with alternative fuels are some of the interesting topics, which are covered in this book.

Keywords Methanol · Biofuels · IC engines · Emissions

Gasoline and mineral diesel are the two main automotive fuels in which gasoline is used for personal transport vehicles (two wheelers and cars) as well as public/commercial transport vehicles (buses, trucks and other light- and heavy-duty vehicles), which are mainly fuelled by diesel. In the last few decades, global transport energy usage increased steadily at a rate of ~ 2 to 2.5% per year, closely paralleling the growth in economic activity globally. International Energy Agency (IEA) scenarios predicted a significant increase ($\sim 50\%$) in global transport energy

A. K. Agarwal (✉) · N. Sharma
Indian Institute of Technology Kanpur, Kanpur, India
e-mail: akag@iitk.ac.in

A. Gautam
RDSO Lucknow, Lucknow, India

A. P. Singh
University of Wisconsin-Madison, Madison, USA

A. K. Agarwal et al. (eds.), *Methanol and the Alternate Fuel Economy*,
Energy, Environment, and Sustainability,
https://doi.org/10.1007/978-981-13-3287-6_1

demand by 2030, which may possibly be further doubled by 2050 (Repowering Transport 2011). Consumption of fossil fuels at such higher rate is creating serious concerns for the human health and the environment. Rapidly increasing fossil fuels consumption leads to a significant rise in fuel prices. These issues have motivated researchers to explore alternative fuels for transportation sector. In the last two decades, a variety of alternative fuels such as biofuels, alcohols, gaseous fuels, biodiesel have been investigated throughout the world. In these alternative fuels, alcohols have shown significant potential for automotive applications. In the alcohol family, methanol, ethanol and butanol can be used in internal combustion (IC) engines; however, ethanol and butanol have several challenging issues such as production limitation and solubility with mineral diesel. These issues limit their application in IC engines. In the last few years, methanol has gained the significant attention of researchers due to its excellent fuel properties, greater compatibility for utilization in existing engines and easy production techniques. Therefore, the first section of this monograph is covering various aspects of methanol.

The first chapter of the monograph describes various methodologies for methanol utilization in IC engines. Important properties such as high octane number, high latent heat of vapourization, zero sulphur content and the presence of oxygen in methanol make it suitable as a fuel for IC engines. This chapter also discusses different techniques for methanol production, in which methanol production from coal and natural gas is the prominent method. Authors presented a list of techniques for methanol utilization in diesel engines. Amongst these techniques, fuel blending and port injection of methanol with diesel pilot injection are the most accepted and used techniques. This chapter also describes the safety aspects of methanol and presents a road map for methanol economy. Methanol utilization strategy is not limited only for road transportation sector. Indian Railways is also desperately looking for the use of methanol as traction fuel. The second chapter of this monograph describes the challenges of Indian Railways and their strategy for methanol utilization for railway traction. This chapter shows that use of methanol in locomotives will provide unparalleled advantages on all three 'E' dimensions, i.e. Economy, Efficiency and Environment. This chapter discusses the road map for efficiency improvement and adoption of methanol as traction fuel for self-powered propulsion for Indian Railways. Third chapter focuses on methanol utilization in rural areas. This chapter shows that methanol can be produced from biomass and it can potentially substitute the petroleum-based products in rural areas. This chapter emphasizes the appropriate strategy development for methanol production and utilization in rural areas. It emerges that an indigenous technology for gasification of biomass should be developed. A consistent policy for the technology development and its market adoption are a few other important criteria, which need to be considered for a success of the plan.

Next section of this monograph is based on different alternative fuels such as biofuels, biodiesel, waste cooking oil and essential oil. First chapter of this section focuses on role of biodiesel in global energy scenario, methods of biodiesel production, fuel properties of biodiesel, the advantages and disadvantages of biodiesel utilization in engines. This chapter clearly indicates that biodiesel improves engine performance and reduces emissions, when used in engines with optimized parameters. Next chapter of this monograph describes the economic factors related to biodiesel production. In this chapter, the authors investigated the sustainability of biodiesel produced from different edible oil crops. They used wheat germ, groundnut and cottonseed oils for biodiesel production using three different processes, namely alkali-catalysed, acid-catalysed and lipase-catalysed transesterification processes. For sustainability analysis, different parameters such as emergy sustainability index (ESI), emergy investment ratio (EIR), emergy renewability (% R), emergy yield ratio (EYR), environmental loading ratio (ELR) and environmental impact ratio (EVR) are evaluated for each of the biodiesel derived from edible oil crops. The authors showed that the biodiesel produced from cottonseed oil is the most sustainable, followed by groundnut oil, and then the biodiesel derived from wheat germ oil crop, which is the least sustainable option amongst the ones considered. In the next chapter, biodiesel produced from waste frying oil has been discussed for engine applications. The authors showed that waste frying oil methyl esters (WFOME) can be effectively used in a blended form with mineral diesel to cater to the energy demands. They suggested that addition of alcohol in the blends of mineral diesel and methyl esters of waste frying oil resulted in improved combustion, performance and emission characteristics. In the last chapter of this section, use of essential oils has been recommended in diesel engines. These essential oils can be produced from the non-fatty parts of the plant (roots, bark, leaves, stems and flowers), and they have similar properties as that of mineral diesel. Due to a low cetane number, essential oils can be used in compression ignition engines by blending with diesel/biodiesel. This chapter shows that use of essential oil in diesel engines results in improved performance and significantly lowers the emissions, especially particulates.

Last section of this monograph describes two different topics related to laser ignition of gaseous fuels and particulate emissions. This chapter shows that compressed natural gas (CNG) is the cheapest alternative fuel with quickest commercial implementation potential due to only marginal modifications required in the existing engine hardware. This chapter suggests that slower flame speed and higher cyclic fluctuations of CNG-fuelled engines can be reduced by hydrogen-enriched CNG (HCNG). Further, use of laser ignition instead of conventional spark ignition system under lean fuel–air mixture condition leads to superior engine performance and lower emissions. Last chapter is based on one of the most serious issues of automotive sector, namely particulate emissions. This chapter describes the effectiveness of various techniques of particulate reduction and suggests the urgent need of switching from the commercially available energy sources to carbon-free fuels. The fuels with a low carbon-to-hydrogen ratio as well as low aromatic content can reduce the particulate emissions substantially.

This monograph presents different aspects such as production, economy and utilization of methanol and other alternative fuels such as biofuels, biodiesel, CNG and essential oils. A unique and important chapter about utilization of methanol in Indian Railways has been also included in this monograph. These topics are organized in four different sections: (i) General, (ii) Methanol Economy,

(iii) Alternative Fuels and (iv) Utilization Aspects. Specific topics covered in the manuscript include:

- Introduction to Methanol and the Alternate Fuel Economy
- Methanol as an Alternative Fuel for Diesel Engines
- Improving Efficiency of Diesel Traction and Adoption of Liquid Sunshine for Indian Railways
- Enabling Rural Economy in India to Partially Substitute Petroleum Products by Methanol—Technology Solutions and Policy Support
- Sustainability Assessment of Biodiesel Production in India from Different Edible Oil Crops Using Emergy Analysis
- Impact of Tri-Fuel on Compression Ignition Engine Emissions: Blends of Waste Frying Oil, Alcohol and Diesel
- Review on the Use of Essential Oils in Compression Ignition Engines
- Study of Performance and Emissions of Engines Fuelled by Biofuels and its Blends
- Laser-Ignited Engine Development for Adaptation to Hydrogen-Enriched Compressed Natural Gas (HCNG)
- Particulate Matter and its Impact on Human Health in Urban Settings.

Reference

Report on Repowering Transport (2011) World Economic Forum, Geneva

Part II
Methanol Economy

Chapter 2
Methanol as an Alternative Fuel for Diesel Engines

Hardikk Valera and Avinash Kumar Agarwal

Abstract Global economic prosperity has led increasing population and a new era of motorization. Petroleum-based reserves are fulfilling the demand for global transport energy; however, petroleum reserves are rather limited and dwindling fast. This alarming situation demands immediate introduction of alternative fuels of bio-origin such as biodiesel, alcohols, vegetable oils. Among various primary alcohols, methanol has emerged as a strong alternate fuel candidate with the highest potential, and it has the potential to significantly contribute to the reduction in crude oil dependence and environmental preservation. Methanol can be straightway used as a replacement for gasoline, since it has very high octane number and has been successfully used in many spark ignition (SI) engine applications. However, utilization of methanol in compression ignition (CI) engines is quite challenging. This chapter deals with many challenges and opportunities of using methanol in CI engine applications.

Keywords Diesel engine · Methanol production · Methanol utilization
Safety

Abbreviations

SI	Spark ignition
CI	Compression ignition
HC	Hydrocarbon
IC	Internal combustion
CNG	Compressed natural gas
MSW	Municipal solid waste
GHG	Greenhouse gas
PM	Particulate matter
BSFC	Brake-specific fuel consumption
DI	Direct injection

H. Valera · A. K. Agarwal (✉)
Engine Research Laboratory, Indian Institute of Technology Kanpur,
Kanpur 208016, India
e-mail: akag@iitk.ac.in

A. K. Agarwal et al. (eds.), *Methanol and the Alternate Fuel Economy*,
Energy, Environment, and Sustainability,
https://doi.org/10.1007/978-981-13-3287-6_2

Nomenclature

H_2	Hydrogen
B_{th}	Brake thermal efficiency
CO	Carbon monoxide
NO_X	Nitrogen oxide

2.1 Introduction

Transport sector has played a vital role to meet mobility requirements of people and goods world over ever since the beginning of industrial revolution. Advancement in the transport sector has improved the human lifestyle leading to better living standards, easy availability of goods of primary needs at a low price, and extended range of goods to be consumed in any geographical region. Specific modes of transport include air transport, sea transport, and land transport. Every transport mode has been powered by specific propulsion system, namely gas turbines, and propeller blades are used for air transport, large-bore internal combustion (IC) engines are used for sea transport, and light-duty/ heavy-duty diesel/gasoline engines are used for land transport. The data for fuel type for engines greater than 2 L displacement reveals that 39% of entire engine produced is diesel-powered. Global diesel engine production will increase from 17.7 million units in 2015 to 21.2 million units by 2021 (Diesel Progress International (January–February 2016). However, nowadays the automobiles powered by diesel engines are facing twin crises. First, a crisis of fuel resource exhaustion due to indiscriminate extraction and extravagant utilization of crude oil (Agarwal 2007). According to an indication, the reserves will keep going for next 114, 50.7, and 52.8 years under reserves-to-product ratio for coal, oil, and natural gas, respectively (BP statistical review of world energy 2016). Second, environmental degradation has led to an increase in global surface temperature (1.1 °C) since the late nineteenth century (https://www.ncdc.noaa.gov/monitoring-references/faq/indicators.php). As a result, the oceans are warming, i.e., increase in temperature (0.302 °F) since 1969 (Levitus et al. 2009). Greenland has already lost 36–60 cubic miles of ice, and Antarctica has lost 36 cubic miles of ice per year between 2002 and 2005 (Tapley et al. 2004). There is an increase in acidity of ocean surface water by ~30% (https://www.pmel.noaa.gov/co2/story/Ocean+Acidification). Exploration of sustainable alternative fuels is one of the important solutions to tackle both the issues. For this, the important attributes of alternative fuels should be:

- They should be produced from the non-petroleum resources.
- They should be eco-friendly.
- They should be available at a low price.
- They should not affect the durability of an engine.

Alcohols, liquefied petroleum gas, vegetable oils, biofuels, natural gas, and hydrogen have been explored by several researchers as alternative fuels. Among these alternate fuel options, alcohols are emerging as the best solutions because they are a part of oxygenate fuel family. They contain hydrocarbons with a hydroxyl group, which contribute to relatively smoother combustion. Primary alcohols can be utilized as fuel in conventional IC engines (Cheung et al. 2008; Hansen et al. 2005; Kisenyi et al. 1994; Kremer et al. 1996). Most preferred alcohols for use in CI engines are methanol, ethanol, propanol, and butanol. These are the simplest primary alcohols, which have comparable auto-ignition temperature, heat of vaporization, and stoichiometric air–fuel ratio as that of conventional mineral diesel. Utilization of these primary alcohols in CI engines offer advantages such as oxygen-enriched test fuels, dominant premixed combustion phase, and improvement in diffusion combustion phase (Lu et al. 2004a, 2004b). Moreover, cleaner burning characteristics of alcohols marginally reduce the emissions of carbon monoxide (CO), hydrocarbons (HC), and oxides of nitrogen (NO_x) (Kim and Dale 2005; Guerrieri et al. 1995; Taylor et al. 1996). Methanol is the simplest alcohols among these primary alcohols, which is a single carbon compound with a hydroxyl group. It contains 50% oxygen by weight. It can emerge as the cleanest alternative fuel for future transport needs if different difficulties could be taken care.

2.2 Properties, Opportunities, and Challenges of Methanol

Methanol has a capability to knock out the conventional fuels such as diesel, compressed natural gas (CNG), and gasoline from the market. It also does not have as severe adverse effects on environment and the human health. Methanol could be used as an alternate fuel in diesel engines; however, it faces several technical challenges, which needs to be overcome.

2.2.1 Properties of Methanol

Methanol is a colorless, volatile, and flammable liquid at room temperature and can be handled as any other conventional liquid fuel. Physicochemical properties of methanol are quite different from other conventional fuels, as shown in Table 2.1 (https://www.mandieselturbo.com/docs/default-source/shopwaredocuments/using-methanol-fuel-in-the-man-b-w-me-lgi-series.pdf?sfvrsn=4).

Combustion Properties Self-ignition temperature is one of the most important factors affecting combustion in a CI engine. The fuel–air mixture gets heated during the ignition delay period in a CI engine. The fuel–air mixture gets auto-ignited after

Table 2.1 Important properties of different fuels (https://www.mandieselturbo.com/docs/default-source/shopwaredocuments/using-methanol-fuel-in-the-man-b-w-me-lgi-series.pdf?sfvrsn=4)

Property	DME	Methanol	Ethanol	Diesel	Gasoline
Chemical formula	CH_3–O–CH_3	CH_3–OH	C_2H_5–OH	C_8–C_{25}	C_4–C_{12}
Fuel carbon (wt%)	52.2	38	52	85	86
Fuel hydrogen (wt%)	13	12	13	15	14
Fuel oxygen (wt%)	34.8	50	35	0	0
Molar mass (kg/kmol)	46	32	46	183	114
Liquid density (kg/m^3)	660	798	794	840	740
Lower heating value (MJ/kg)	22.8	20.1	27.0	42.7	–
Boiling temperature (°C at 1 bar)	−24.9	65	78	180–360	27–245
Vapor pressure (bar at 20 °C)	5.3	0.13	0.059	«1	0.25–0.45
Critical pressure (bar)	53.7	81	63	30	–
Critical temperature (°C)	127	239.4	241	435	–
Kinematic viscosity (cSt at 20 °C)	0.19–0.25	0.74	1.2	2.5–3.0	0.6
Surface tension (N/m at 20 °C)	0.012	0.023	0.022	0.027	–
Bulk modulus (N/mm^2 at 20 °C 2 MPa)	1549	823	902	553	1300
Cetane number	55	<5	8	38–53	–
Octane number	low	109	109	15–25	90–100
Auto-ignition temperature in air (°C)	350	470	362	250–450	250–460
Heat of vaporization (kJ/kg at 1 bar)	467	1089	841	250	375
Minimum ignition energy (mJ at $\varphi = 1$)	0.33	0.21	0.65	0.23	0.8
Stoichiometric air/fuel ratio	9	6.5	9.1	14.6	14.7
Peak flame temperature (°C at 1 bar)	1780	1890	1920	2054	2030
Flammability limits (vol.%)	3.4–18.28	6–36	3–19	0.5–7.5	1.4–7.6
Flash point (°C)	−41	12	14	52	–45

the ignition delay period, post-attainment of the auto-ignition temperature in the combustion chamber. Methanol has higher auto-ignition temperature compared to baseline mineral diesel (methanol: 464 °C, and mineral diesel: 240 °C); hence, it is a safer fuel for transportation from one place to another. Also, dedicated methanol engines can be designed to operate smoothly at a higher compression ratio; hence, they can be more efficient thermodynamically.

Molecular Weight Methanol has a molecular weight of 32, which is approximately six times lighter than mineral diesel. It results in a lower emission because the diffusion rate for lighter fuel is lower compared to heavier fuel, i.e., petroleum-based fuels.

Oxygenated Fuel Methanol has inherent oxygen in its molecular structure, which is responsible for reduction of emission of CO and NO_x by converting into a carbon dioxide (CO_2) and NO_2. Further, it helps in achieving more complete combustion during the expansion stroke.

Latent Heat of Vaporization Methanol has higher latent heat of vaporization than conventional fuels. It therefore provides extra cooling effect to the intake charge compared to petroleum-based conventional fuels. Hence, it improves the brake thermal efficiency and power output.

Sulfur Content Methanol contains zero sulfur; hence, its use in the engines results in zero emission of sulfur-based pollutants, i.e., SO_2 and SO_3, which are mainly responsible for acid rain.

Heating Value Methanol has lower calorific value than baseline mineral diesel due to inherent oxygen in its molecular structure. Therefore, higher fuel quantity is required to be injected in order to achieve an equivalent brake power output as that of a diesel-powered engine. Hence, larger methanol storage tank is required to achieve the same range of the vehicle, which draws the same power from it.

Cetane Number Cetane number of methanol is <5, which is very low compared to mineral diesel. This characteristic of methanol creates problems in obtaining smooth combustion similar to petroleum-based fuels.

Lubricity and Viscosity Methanol has poor lubricity properties and is also less viscous compared to baseline mineral diesel. This creates issues in the fuel delivery system, plungers, and feed pump and corrodes these components of the fuel injection equipment. Viscosity of methanol is relatively lower than the prescribed ASTM standard range for mineral diesel (1.39–4.2 cst at 40 °C); therefore, suitable additives have to be added to methanol, or it has to be blended with mineral diesel so that it can become a suitable fuel for CI engines.

2.2.2 *Opportunities*

Methanol as a fuel offers several opportunities. These are:

- Methanol can be produced from renewable resources and abundantly available high ash coal, MSW, and low-value biomass.
- Methanol-powered engine produces relatively lower emissions compared to baseline mineral diesel fuelled engines.
- Combustion noise of methanol-powered engines is relatively lesser than equivalent diesel engines.

- Methanol increases the thermal efficiency of engine due to charge cooling and fuel oxygen. This leads to higher well-to-wheel efficiency vis-à-vis baseline mineral diesel.
- It offers higher land use efficiency than any other cultivable renewable fuel.

2.2.3 Technical Challenges and Potential Solutions

- Methanol has less energy density per unit mass, i.e., 23 MJ/kg, which is nearly half than diesel which necessitates the modification in existing diesel fuel injector, i.e., upsizing of nozzle's diameter to get equivalent brake power output as diesel can.
- Methanol has low viscosity and lubricity which results in leakage of it from O-rings, gaskets, gland seals, and packings and unexpected, catastrophic containment failure of sealed joints. Among various polymers, fluoroethylene propylene-perfluoroalkoxy has an excellent chemical resistance which can help to avoid unnecessary leakage of methanol from a diesel engine.
- Introduction of the methanol–diesel blend has a noticeable effect on the NO_x concentration at high engine load, whereas it has a little effect at low engine load. This effect can be optimized by sending the calculated amount of air through open ECU such that exhaust temperature should not be increased.

2.3 Methanol Production

Methanol is used as a fuel and as a feedstock for production of pesticides, medicines, formal aldehyde, acetic acid, and dimethyl ether. Figure 2.1 (https://ihsmarkit.com/products/chemical-market-daily-service.html) shows the global demand for methanol and trends in the near future. It is clearly evident from this data that demand for methanol is increasing continuously and has become more than double between 2009 and 2017.

To meet huge demand from chemical sector, methanol is produced from several carbonaceous feedstocks such as natural gas, coal, biomass, and CO_2. Use of low-value biomass and municipal solid waste can resolve triple problems in one go, namely energy demand, waste management, and greenhouse gas (GHG) emissions by using these waste materials as a feedstock and converting them into various forms, using biochemical/thermochemical processes. Later path is preferred compared to biochemical path because of vast availability of feedstock and faster conversion rates (Shahbaz et al. 2017; Guan et al. 2016). The reactions and technology involved in the production of methanol are discussed in detail in the following subsections.

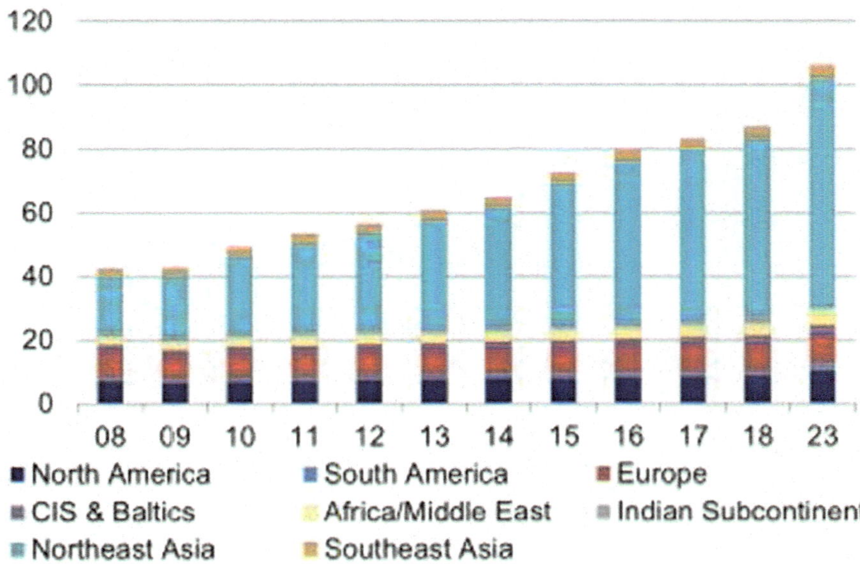

Fig. 2.1 World methanol demand for different geographical regions (https://ihsmarkit.com/products/chemical-market-daily-service.html)

2.3.1 Methanol Production from Coal

This method generally involves four steps: synthetic gas (syngas) generation, syngas purification, methanol synthesis, and methanol rectification. A flow diagram for the production of methanol from coal is shown in Fig. 2.2. Coal slurry is supplied with oxygen under high pressure to the gasifier for generating syngas, which is a mixture of hydrogen (H_2), carbon monoxide (CO), carbon dioxide (CO_2), and hydrogen sulfide (H_2S). Therefore, purification step is required for methanol synthesis, where CO and H_2S are removed and recovered using syngas scrubbing with the help of amine solution. Recovered H_2S is then converted into sulfur. Produced syngas possesses a low ratio of hydrogen to carbon, which is usually not sufficient for methanol synthesis. Hence, shift reaction is essential, wherein the ratio of H_2/CO is maintained for achieving stoichiometry. This process produces additional CO_2 as well, which is removed by scrubbing using amine solution. Finally, to produce purified methanol, shifted gas mixes with a remainder of the syngas and methanol production reaction is performed on zinc–chromium (Zn–Cr) catalyst at 30–35 MPa pressure and 300–400 °C reaction temperature, as shown in Eqs. (2.1)–(2.3).

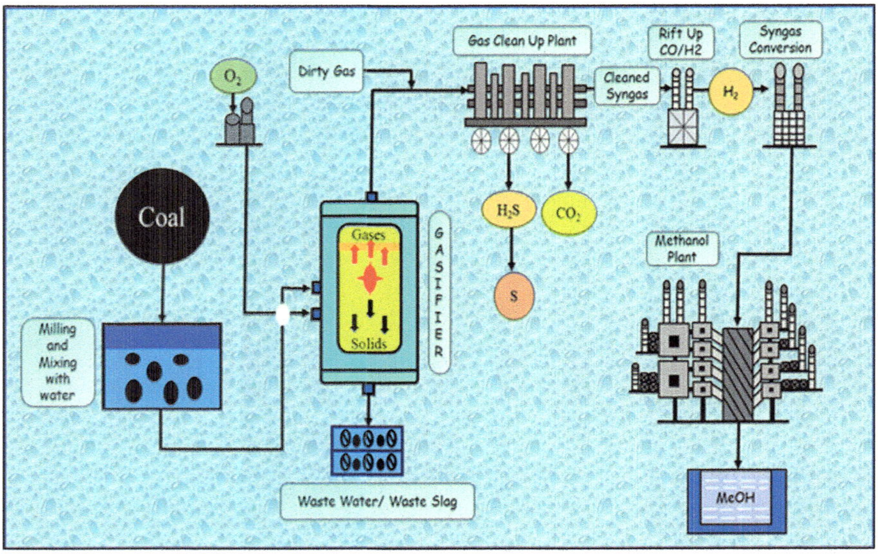

Fig. 2.2 Process flow diagram for coal-to-methanol production

Methanol synthesis reaction:

$$2H_2 + CO \rightarrow CH_3OH \tag{2.1}$$

Water-gas shift (WGS) reaction:

$$CO + H_2O \rightarrow CO_2 + H_2 \tag{2.2}$$

Methanol synthesis reaction:

$$3H_2 + CO_2 \rightarrow CH_3OH + H_2O \tag{2.3}$$

2.3.2 Methanol Production from Natural Gas

The entire process for production of methanol from natural gas involves three necessary steps (Tijm et al. 2001; Roan et al. 2004; Fiedler et al. 2000): production of syngas, conversion of produced syngas into crude methanol, and the distillation of crude methanol to achieve desired purity. Steam reforming and auto-thermal reforming of natural gas is an essential process to produce syngas as shown in Eqs. (2.4)–(2.6). However, syngas can also be produced by partial oxidation of methane as shown in Eq. (2.7).

Steam reforming reaction:

$$CH_4 + H_2O \leftrightarrow CO + 3H_2 \quad (2.4)$$

Auto-thermal reforming (ATR) reaction

$$CH_4 + 2O_2 \rightarrow CO_2 + 2H_2O \quad (2.5)$$

It further involves water-gas shift (WGS) reaction:

$$CO_2 + H_2 \leftrightarrow CO + H_2O \quad (2.6)$$

Partial oxidation reaction:

$$CH_4 + \frac{1}{2}O_2 \rightarrow CO + 2H_2 \quad (2.7)$$

Composition of syngas is expressed by the stoichiometric number S_N, which is the ratio of the two quantities, namely: (1) The difference between number of moles of hydrogen and carbon dioxide, and (2) the summation of the number of moles of CO_2 and CO (Bozzano and Manenti 2016).

$$S_N = \frac{moles\ of\ Hydrogen - moles\ of\ carbon\ dioxide}{moles\ of\ carbon\ dioxide + moles\ of\ carbon\ monoxide}$$

For production of methanol, S_N should be a value of 2. However, S_N depends on the feedstock material used for production of syngas. When syngas is produced from natural gas reforming, S_N usually achieved is 2.8–3. S_N takes into account the presence of CO_2 converted, which consumes hydrogen via the reverse WGS reaction and produces methanol as shown in Eq. 2.8.

$$CO_2 + 3H_2 \leftrightarrow CH_3OH + H_2O \quad (2.8)$$

At high-pressure (50–100 bar) and high-temperature (200–300 °C) conditions, syngas is converted into crude methanol. The main reactions involved in methanol synthesis are shown in Eqs. 2.9–2.12 (Klier 1982):

Hydrogenation of carbon monoxide:

$$CO + 2H_2 \leftrightarrow CH_3OH \quad (2.9)$$

Divided into two consecutive steps:

$$CO + H_2 \leftrightarrow CH_2O \quad (2.10)$$

$$CH_2O + H_2 \leftrightarrow CH_3OH \quad (2.11)$$

Hydrogenation of carbon dioxide:

$$CO_2 + 3H_2 \leftrightarrow CH_3OH + H_2O \tag{2.12}$$

The reactions (2.9) and (2.12) are exothermic, and they consequently require significant cooling. Furthermore, with a ratio of H_2/CO of 3:1–5:1, mixed gases are fed for gas phase processes. Further process reactions also involve equipment, where WGS is used to improve the hydrogen concentration. Also, the conversion of syngas is subjected to a thermodynamic equilibrium which confines this reaction to be a low conversion process. This leads to a large recycling of unconverted gas. The resulting recycling and cooling are the main responsible factor for high investment costs for this process segment.

2.3.3 Methanol Production from Biomass/Municipal Solid Waste

Low-value agricultural waste/municipal solid waste (MSW) are important feedstock resources that remain untapped worldwide. Some landfills have installed gas recovery and waste-to-energy conversion systems via combustion. However, these routes for methanol production utilizing biomass/ MSW generally yield low efficiency. The process for producing methanol from biomass/ MSW is similar to that from natural gas. Entire process is divided into three steps: production of syngas, synthesis of crude methanol, and finally purification. A flow diagram for conversion of grown biomass to methanol is shown in Fig. 2.3. The feedstock is first shredded and then it undergoes a process known as "removal of ferrous materials", wherein ferrous materials are removed with the help of a magnet. After that, gasification process is initiated, which is a thermochemical conversion process to convert solid biomass/ MSW into gaseous form with the help of gasifying agents such as oxygen, steam, and flue gases (Zhao et al. 2015; Dai et al. 2015). It demands high temperatures in presence of oxygen, mainly for large-scale production of methanol. Gas scrubbing is the next process to remove water. Thereafter, other processes are accomplished similarly as in case of the process of methanol production from coal.

2.3.4 Methanol from Catalytic Hydrogenation of CO_2

CO_2 is a very stable molecule. It requires substantial energy input, optimized reaction conditions, and a catalyst with high stability to make it reactive for production of methanol. Bonds between carbon and oxygen are firm and strong; hence, high energy and presence of good catalytic system are essential to break them. The reaction to produce methanol from CO_2 by hydrogenation is given in Eq. 2.13:

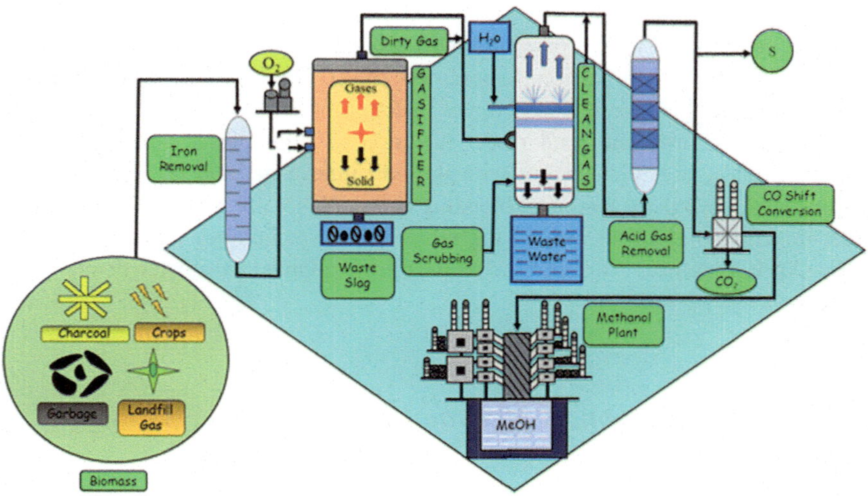

Fig. 2.3 Production flow diagram from biomass/MSW to methanol

$$CO_2 + 3H_2 \leftrightarrow CH_3OH + H_2O \tag{2.13}$$

This method is a practical approach to utilize at CO_2 and reduce its concentration in the atmosphere, which was 403.3 ppm in 2016, up from 400 ppm in 2015 (https://public.wmo.int/en/resources/library/wmo-greenhouse-gas-bulletin). Moreover by this concept, CO_2 can be recovered from an industrial source and human activities and then converted into methanol. Utilization of CO_2 is advantageous because it is inexpensive, abundant, non-toxic, non-corrosive, and nonflammable gas; hence, it is safe to use. Further, it provides ease to store and transport methanol from synthesis plants (Centi and Perathoner 2009). Hydrogen can be extracted from water by electrolysis, as shown in Eq. 2.14, using a renewable source of electricity and used as a reactant for this process.

$$2H_2O \rightarrow 2H_2 + O_2 \tag{2.14}$$

2.4 Use of Methanol in Diesel Engines

Exponential increase in emissions is compelling scientists to either introduce a new environment-friendly fuel for diesel engines or redesign existing engines to meet strict emission legislation. Methanol is a strong contender as an alternative fuel because it emits fewer harmful atmospheric pollutants. Methanol has been widely investigated for application in blended form with diesel, in order to reduce pollutants (Ajav et al. 1998; Saravanan et al. 2002). Moreover, it can also be applied as a non-fossil fuel for reduction in harmful pollutants and for increasing the thermal efficiency.

California program demonstrated 600 vehicles, fuelled by alcohol. Volkswagen and Ford participated in this program and introduced 19 vehicles, which operated using methanol (Jet Propulsion Laboratory 1983). Various techniques to inject methanol in diesel engine included direct injection of methanol–diesel blends, port fuel injection of methanol and direct injection of pilot diesel, methanol–diesel emulsified fuels, use of methanol with ignition improvers, and use of glow plug concept. Out of all these methods, blending and port fuel injection of methanol and direct injection of pilot diesel are the most reliable techniques for methanol utilization in IC engines. All these techniques are explained in greater detail in the following subsections.

2.4.1 Direct Injection of Methanol Blends

Blending is the first choice for any researcher for the introduction of an alternative fuel because no additional device is required for the fuel injection system. In fact, blending is an excellent method to reduce emissions from IC engines. In this method, methanol and diesel are mixed in desired proportion in a fuel tank, and the stable mixture is then injected into the engine combustion chamber via the fuel injector, without any hardware changes in the fuel injection system. This method limits the introduction of large methanol quantity because of its poor miscibility with diesel. Methanol separates out from diesel in proportions larger than 10% (v/v). In order to overcome the phase separation issues, different additives such as co-solvents or binders are introduced in methanol–diesel blends to ensure adequate miscibility of the constituents. Binders eventually minimize the fuel supply to the engine. Moreover, physicochemical properties of methanol and diesel are not similar; hence, blending of two such fuels may lead to changes in essential fuel properties such as fuel viscosity, density, and cetane number. Several studies concluded that alcohol–diesel blends are superior to baseline diesel, leading to lower emissions and higher efficiency (Agarwal et al. 2016; Campos-Fernández et al. 2012). Several studies, which investigated methanol–diesel blends vis-à-vis baseline mineral diesel are summarized in Table 2.2.

The tabulated experimental analysis demonstrates that methanol–diesel blends can be used as fuel in diesel engines. Brake thermal efficiency of methanol blends was higher because of its lower calorific value, whereas the power output and torque for diesel were lower vis-à-vis methanol–diesel blends. The tailpipe emission results varied from researcher to researcher, and there was no unique trend in results. Reduction in CO and HC leads to increase in NO_X emissions, whereas reduction in NO_X leads to increase in CO and HC emissions. Similar trend of pollutants was noticed with methanol–biodiesel blends.

Table 2.2 Important experimental results of methanol–diesel blends

Year	Researcher	Fuel used	Concluding remarks
2001	Chao et al. (2011)	D100 D95M5 D92M8 D90M10 D85M15	NO_X (↓), CO (↑), and (HC) (↑) with increase in methanol percentage in the blend
2003	Kumar et al. (2003)	D100 BD100 D70M30 BD70M30	B_{th} (↑) for D70M30, NO (↓), HC (↓), CO (↓), and smoke (↓)
2004	Huang et al (2004)	D100 D80M20 D60M40 D40M60	Isobutanol and oleic acid used to stabilize diesel–methanol blend. CO (↓) and NO_X (↑) with an increasing mass fraction of methanol in test blend
2008	Cheng et al (2008)	D100 BD100 D90M10 (Fumigation) BD90M10 (Emulsion)	CO_2 (↓): 2.5% compared to baseline fuel emission and NO_X (↓): 5% with BD90M10
2009	Canakci et al. (2009)	D100 D95M5 D90M10 D85M15	**Injection pressure at 180 bar** NO_X (↓), CO (↑), HC (↑), and smoke (↑) with increasing mass fraction of methanol, while other parameters remain unchanged **Injection pressure at 200 bar** Heat release rate (↓), maximum cylinder pressure (↓), fuel conversion efficiency (↓), smoke (↓), HC (↓), CO (↓), BSFC (↑) and NO_X (↑) with an increasing mass fraction of methanol **Injection pressure at 220 bar** CO (↓), smoke (↓) and HC (↓), heat release rate (↑), maximum cylinder pressure (↑) and NO_X (↑) with increasing mass fraction of methanol Increase in NO_X and decrease in CO and UHC emissions because of oxygen content of methanol. Extra oxygen is responsible for converting CO into the CO_2 and UHC into the completely burn hydrocarbon. In addition, improved combustion efficiency is responsible for increasing NO*x* emissions with increasing injection pressure. Peak cylinder pressure was higher for methanol blend compared to diesel because of higher ignition delay
2010	Qi et al. (2010)	D50BD50 D47.5BD47.5M5 D43.2BD43.2M9.6O4 Biodiesel (Methyl Soyate)	At 1500 rpm, similar peak cylinder pressure and peak of pressure rise were found for D47.5BD47.5M5 and D43.2BD43.2M9.6O4 compared to D50BD50 and higher peak heat release rate compared to D50BD50. At 1800 rpm, peak cylinder pressure and peak pressure rise for D47.5BD47.5M5 and D43.2BD43.2M9.6O4 were lower compared to

(continued)

Table 2.2 (continued)

Year	Researcher	Fuel used	Concluding remarks
			D50BD50. Furthermore, CO emissions from D47.5BD47.5M5 and D43.2BD43.2M9.6O4 were lower compared to B50D50.t speed of 1500 rpm, similar peak cylinder pressure and peak of pressure rise are found for D47.5BD47.5M5 and D43.2BD43.2M9.6O4 to D50BD50 and higher peak of heat release rate compared to D50BD50. At speed of 1800 rpm, peak cylinder pressure and peak of pressure rise for D47.5BD47.5M5 and D43.2BD43.2M9.6O4 are lower compared to D50BD50. Furthermore, CO emissions of D47.5BD47.5M5 and D43.2BD43.2M9.6O4 are lower compared to B50D50
2010	Zhu et al. (2010)	D100 BD100 BD95E5 BD90E10 BD85E15 BD95M5 BD90M10 BD85M15 Biodiesel (Waste Cooking oil)	HC (↓): 10%, CO (↓): 15% and NO_X (↑) at 5% blends BSFC (↓), PM (↓) compared to diesel
2011	Ciniviz et al. (2011)	D100 D95M5 D90M10 D85M15	NO (↑), CO (↓), HC (↓)
2014	Yasin et al. (2014)	D100 D80BD20 D75BD20M5	B.P. (↓), CO (↓), CO_2 (↓), (BSFC) (↑), and NO*x* (↑): 13%
2016	Agarwal et al. (2016)	D100 D95BD5 D80BD20 D95M5 Biodiesel (Karanja)	**Regulated emissions** CO (↑) at higher engine loads, HC (↓) decreased at higher engine loads, HC (↑) at lower engine loads, and NO*x* (↓) **Unregulated emissions** Methane (↑), n-pentane (↑), acetylene (↑), n-butane (↓), formaldehyde (↓), acetaldehyde (↓), and ethanol (↓) with increasing engine load
2016	Bharadwaz et al. (2016)	BD100 BD95M5 BD90M10 BD85M15 Biodiesel (Pure Palm Oil)	B_{th} (↑), BSFC (↑), CO (↓), HC (↓), NO (↓), and smoke (↓)

(continued)

Table 2.2 (continued)

Year	Researcher	Fuel used	Concluding remarks
2016	Prashant et al. (2016)	D100 D80M20 D60M40 D40M60	Maximum pressure rise increased due to rise in ignition delay with methanol compared to diesel. Maximum heat release was noticed at D60M40

D diesel, *M* methanol, *BD* biodiesel, *DO* dodecanol, *O* oleic acid, *E* ethanol and DaMbBDcEd: Diesel a% (v/v), Methanol b% (v/v), Biodiesel c% (v/v), and Ethanol d% (v/v)

2.4.2 *Port Fuel Injection of Methanol and Direct Injection of Pilot Diesel*

In "port fuel injection of methanol and direct injection of pilot diesel" method, methanol is injected into the intake air stream and diesel pilot is injected directly into the engine combustion chamber. Port-injected methanol burns and became a contributor to the power produced. This process is divided into major fraction and minor fraction port fuel injection. When more than 50% methanol injection in the port is used, it is known as a major fraction port fuel injection. Minor fraction port fuel injection is under 50% methanol injection in the port. Port fuel injection of methanol and direct injection of pilot diesel is an expeditious way of introducing methanol into diesel engines. It helps achieve smoother engine operation and permits significantly higher quantity of methanol to be injected in the engine. A schematic of this technology is shown in Fig. 2.4. A low-pressure injector and a high-pressure injector are used to inject methanol and diesel, respectively. It also requires a control system to control the methanol and diesel injection quantity and timing, etc. However, this method increases the net weight of the system because it requires additional fuel injection equipment for the introduction of methanol (Abedin et al. 2016). This approach is therefore one of the most competitive approaches for the introduction of methanol into the engine combustion chamber. Several researchers have investigated this technology, and the results of some of these studies are summarized in Table 2.3.

2.4.3 *Methanol Emulsions*

The very limited solubility of methanol in diesel has led to extensive research for finding ways and means for using methanol via emulsion route. By using emulsions, it was possible to add a larger amount of methanol to diesel (tests using 10–30% were quite common), compared to what was possible in blending route. Hence, several researchers investigated this method, and their main research findings are summarized in Table 2.4.

Several researchers tried using methanol emulsions in the IC engines. Some important problems associated with utilization of methanol via emulsions technique are as follows:

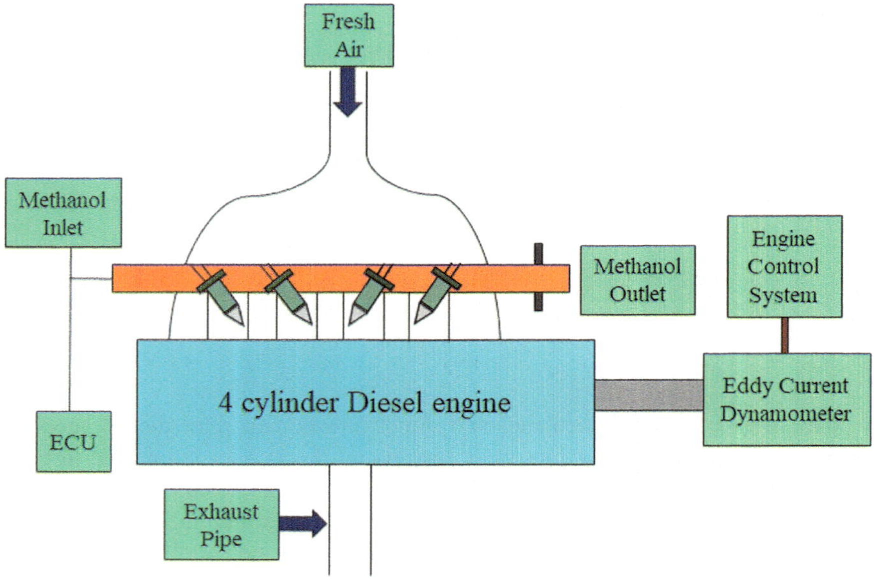

Fig. 2.4 Port fuel injection of methanol and direct injection of pilot diesel

Table 2.3 Important experimental findings of port injection of methanol and direct injection of pilot diesel technique

Year	Researcher	Test fuel	Remarks
1980	Houser et al. (1980)	D60M40	Approach is helpful in reducing NO_X emissions. Smoke opacity (↑) and fuel efficiency (↑) at higher loads
1981	Heisey and Lestz (1981)	–	B_{th} (↑) at moderate, heavy loads, ignition delay (↑) at all loads and CO (↑) with fumigation
1992	Odaca et al. (1989)	–	Total fuel consumption (↑) and smoke (↓) with an increase in methanol energy ratio
2008	Yao et al. (2008)	D100 BD100 D70M30 BD70M30	NO_X (↓), smoke opacity (↓): 50% compared to conventional diesel and smoke opacity (↓) with an oxidation catalytic converter
2011	Zhang et al. (2011)	D100 D80M20 D60M40 D40M60	BSFC (↑), PM (↓), NO_X (↓), CO (↑) and HC (↑) with an increase in fumigation ratio
2013	Zhang et al. (2013)	D100 BD100 D90M10 (Fumigation) BD90M10 (Emulsion)	This method increases the peak heat release rate and ignition delay

Table 2.4 Important experimental findings of using methanol emulsions technique

Year	Researcher	Fuel used	Concluding remarks
2008	Bayraktar (2008)	D100 D94M05DO1 D89M10DO1 D84M15DO1	Performance wise 10% blend of methanol with diesel is a promising solution. They showed 7% improvement in performance while using diesel–methanol emulsion with dodecanol as an additive
2010	Sayin. (2010)	D100 D94M05DO1 D89M10DO1 D94E05DO1 D89E10DO1	Dodecanol used as an additive to overcome the phase separation problem Smoke (↓), CO (↓), HC (↓), NO_X (↑), BSFC (↑), and fuel conversion efficiency (↓)

- Methanol requires large concentration of emulsifier, which is not economically viable.
- Injection of methanol reduces the cetane number of the test fuel, which necessitate changes in the fuel injection timing of the engine.
- Emulsions become quite viscous at low temperatures and tend to separate in the presence of water, thus creating low-temperature stability issues.
- Energy density of emulsions is lower than baseline diesel, which necessitates adjustments in fuel injection system in order to maintain the power output.

2.4.4 Use of Methanol with Ignition Improvers

Methanol has a very low cetane number. This is attributed to relatively higher ignition delay period compared to baseline mineral diesel. Furthermore, it leads to a higher rate of pressure rise, leading to uncontrolled knocking. Ignition improvers, also are known as cetane improvers, are a promising solution to overcome the knocking problem of methanol-fuelled engines. Cetane improver is essentially an additive, which decomposes at lower temperature, and their exothermic decomposition leads to low-temperature combustion. It permits the use of methanol in a diesel engine without costly hardware modifications. Also, it provides flexibility to the drivers for switching between the fuels. Diethyl ether and alkyl nitrates can be used as cetane improver for methanol-fuelled engines.

Historical seeds of ignition improver deployment are in Brazil. In 1979, Mercedes-Benz demonstrated it on buses with ethanol as the main fuel. They converted hundreds of trucks to fuel them with the help of ignition improvers. However, most ignition improvers contain nitrogen in their molecular structure; therefore, they increase NO_X emissions. Testing showed an increased level of NO_X; however, overall NO_X emissions decreased. This was attributed to superior combustion of methanol.

2.4.5 *Glow Plug Concept*

During cold weather conditions, methanol suffers from the poor ignition characteristics and slower flame propagation. Glow plug is a promising solution to overcome cold start problems. Glow plug has a heating element, which ignited methanol for smoother combustion. Detroit Diesel Corporation used the concept of glow plug for the compression ignition version of their two-stroke diesel engine, which was practically demonstrated and used by hundreds of buses and other heavy-duty vehicles. This engine used a glow plug to achieve compression ignition at low engine loads and used heat from burned gasses to heat up methanol at high engine loads. They succeeded in reducing NO_X emissions, but at the same time, PM emissions increased due to consumption of lubricating oil. Caterpillar developed a methanol-fuelled four-stroke diesel engine with the help of a glow plug in order to achieve appropriate ignition (Richards 1990). Navistar also successfully developed a DT- 466, four-stroke diesel engine using glow plug concept (Baranescu et al. 1989). Suresh et al. (2010) conducted engine experiments with and without the glow plug. Without glow plug, they obtained poor brake thermal efficiency, higher hydrocarbons, and carbon monoxide emissions vis-à-vis baseline mineral diesel. Subsequently, they repeated the experiments using glow plug, and their experimented results showed improved thermal efficiency by ~3% and reduction in hydrocarbons, carbon monoxide, and smoke emissions by ~69%, ~50%, and ~9%, respectively. The presence of a glow plug did not affect NO_x emissions.

A comparison of the above-mentioned techniques on the basis of experimental findings is summarized and tabulated in Table 2.5.

2.5 Safety of Methanol-Fuelled Engine

Physical properties of the fuel play a vital role in estimating the probability of a fire hazard, including the extent of fire. Fuel properties such as volatility, self-ignition temperature, and flammability range affect the fire hazards, energy density, flame temperature, and thermal radiation, which in turn show the potential risk associated in an event of accidental fuel fire. Estimation of the fire risk can be calculated by considering two situations: (1) Fuel is exposed to open atmosphere, and (2) fuel is contained in an enclosed tank. Important factors include fuel volatility, flammability limits, fuel vapor density, diffusion coefficient, and source of ignition. Key factors for fuel risk assessment are shown in Table 2.6. The detailed discussion on these risk factors is given in the following paragraphs.

Table 2.5 Comparison of various methanol induction techniques in the engine

Fuel induction method		Direct injection of methanol blends	Port fuel injection of methanol and direct injection of pilot diesel	Methanol emulsions	Methanol with ignition improvers	Glow plug concept
		Direct injection	Methanol is injected into the intake air and diesel direct injection in the combustion chamber	Direct injection	Direct injection	Direct injection
Effect on performance	B_{th}	(↑)	(↑)	(↑)	–	(↑)
	BSFC	(↑)	(↑)	(↓)	–	–
Effect on emissions	CO	(↓)	(↑)	(↓)	–	(↓)
	HC	(↓)	(↑)	(↓)	–	(↓)
	NO_X	(↑)	(↓)	(↑)	(↓)	–
	Smoke	(↓)	(↓)	–	–	(↓)

Table 2.6 Fire risk assessment properties

Situations	Key factors
Open to atmosphere	• Fuel vapor density • Diffusion coefficient • Source of ignition
In the tank	• Fuel volatility • Flammability limits • Ignition properties

2.5.1 *Fuel Exposed to the Open Atmosphere*

Diesel Diesel is hardly ignitable in open atmosphere because of its extremely low volatility in ambient conditions. However, combustible vapor mixture forms only at temperature above 56 °C. Furthermore, diesel has low auto-ignition temperature, which requires an ignition source.

Methanol Methanol is relatively harder to ignite in open atmosphere. Basic reasons behind this are mentioned below:

- Low vapor pressure.
- Low vapor density and high diffusion coefficient.
- High self-ignition temperature.

Table 2.7 Summary of accidental fire-related issues of diesel and methanol

In case of a fire	Diesel	Methanol
Difficulty	Fire of mineral diesel starts slowly. But, if it starts, then it progresses rapidly and violently. It generates dangerous smoke	Methanol has a very high latent heat of vaporization that allows a slower and controlled fire, with no flame radiations
Visibility	Fire of diesel will generate yellow flames, which are visible under any conditions	Methanol flames have very low visibility. Additives should be used with methanol to improve the flame visibility
Solution to avoid a fire	CO_2 is the best solution to fight diesel fire	Methanol is soluble in water; therefore, it is not a suitable solution to arrest methanol fire. Methanol resistant fire extinction foams are the best solution to fight with methanol fire

2.5.2 *Fuel Contained in an Enclosed Tank*

Diesel Fire risk from mineral diesel is quite low because it requires at least 56 °C temperature for formation of ignitable combustible vapor mixture. However, return fuel from a firing engine increases the fuel temperature in the tank. But this situation is unlikely to lead to a fire incident since most modern vehicles are equipped with fuel coolers.

Methanol Fire risk from methanol is relatively higher because it needs only 12 °C temperature to form combustible vapor mixture. For this reason, associated fire risk with return fuel from a firing engine is relatively higher.
Recommendation to avoid fire risk are as follows:

- Fuel tank should be made of anti-corrosive material.
- Fuel refill pipe should be made of warm galvanized steel.
- Fuel distribution line should be made of a plastic.
- Flame arrester must be used at refill systems of the enclosed tank.

Some important aspects related to accidental fire from engine fuels, flame visibility, and potential solution are summarized in Table 2.7.

2.6 Action Plan for Developing Methanol-Fuelled Diesel Engines

Methanol is projected to be one of the best alternate energy carriers for diesel engines in foreseeable future. However, there is a need for several technological interventions. An exhaustive study needs to be conducted for commercialization of such methanol-fuelled engines. Specific action points for developing methanol-fuelled diesel engines that need to be investigated comprehensively are as follows:

- Development of methanol compatible fuel injection system,
- Development of optimized combustion strategy,
- Emission development and certification,
- Component refinement and cost reduction,
- Trials of methanol-fuelled engines and optimization of systems and components,
- Durability studies of the fuel injection equipment and engine systems,
- Reliability and safety demonstration,
- Development of methanol production capacity,
- Creation of infrastructure for methanol distribution infrastructure,
- Fuel certification and protocols.

2.7 Conclusions

A comprehensive review of methanol production through renewable as well as non-renewable feedstocks has been done in this chapter. Methanol utilization in compression ignition engine vehicles is also discussed comprehensively. Methanol is one of final frontiers for green chemistry, both regarding its large-scale production and its utilization in diesel engines, leading to rise of methanol economy. From a production point of view, there are two important environment-friendly routes: (i) methanol production from MSW/biomass and (ii) catalytic hydrogenation of atmospheric CO_2 for methanol production, which has a great potential to reduce GHG emissions. In the first route, GHG is reduced by producing energy from waste. This would also play a mammoth role in resolving the problem of waste disposal. In the second route, it will be possible to reduce CO_2 emissions by capturing it from any industrial source, human activity, or environment by absorption and then chemically converting it into methanol. Production of methanol from CO_2 is regarded as noteworthy because it is a green process, considering that H_2 required for the process can be produced by electrolysis of water using any renewable source of primary energy. The table below summarizes various aspects of methanol economy, which have been discussed in this chapter.

Alternative fuel options for diesel engines	• There is a growing concern for the environmental pollution from diesel engines worldwide. World required non-petroleum-based alternative fuels for diesel engines • Resources, acceptability, effect of pollutant on the environment, technology, and versatility could be some of the deciding factors for the introduction of alternative fuel for diesel engine • Direct injection of methanol blends, port fuel injection of methanol and direct injection of pilot diesel, use of methanol with ignition improvers, and use of glow plug are some of the proposed methods to introduce methanol in diesel engines
Production of methanol	• Methanol production from coal, natural gas, MSW and catalytic hydrogenation of CO_2 are some of the Industrial scale methods to cater to huge demand for methanol • Methanol from renewable sources such as MSW and catalytic hydrogenation of CO_2 could address the twin crises: depletion of fossil fuels and environment degradation
Use of methanol in diesel engine	• Existing diesel engines do not allow the use of 100% methanol due to cold start and durability issues • This requires exploration of plausible routes to introduce methanol up to the extent possible. Direct injection of methanol blend, methanol fumigation, and emulsions are the expeditious ways for utilization of methanol in diesel engine • The cold start issues can be overcome by using glow plug • Higher ignition delay of methanol creates knocking issues due to its low cetane number. Ignition improvers are a promising solution to overcome knocking problem. However, it is an expensive method
Combustion characteristics of methanol	• For same energy input, methanol-powered engine exhibits superior engine performance and emissions characteristics compared to baseline mineral diesel-powered engines
Safety of methanol for diesel engines	• Methanol is a colorless, tasteless, volatile, and flammable liquid. Accidental drinking should be avoided because it is poisonous and affects the nervous system. Even a small exposure may result in visual deficiency, coma, and demise. Ingestion of 10 ml can cause blindness, and 60–100 ml can be fatal

Finally, methanol-fuelled engines are capable of meeting Euro-6 emission legislations, which will be implemented in India by 2020; hence, the introduction of methanol economy in the country makes a lot of sense and should be considered seriously.

References

Abedin MJ, Imrann A, Masjuki HH, Kalam MA, Shahir SA, Varman M, Ruhul AM (2016) An overview on comparative engine performance and emission characteristics of different techniques involved in diesel engine as dual-fuel engine operation. Renew Sustain Energy Rev 60:306–316

Agarwal AK (2007) Biofuels (alcohols and biodiesel) applications as fuels for internal combustion engines. Prog Energy Combust Sci 33:233–271

Agarwal AK, Shukla PC, Patel C, Gupta JG, Sharma N, Prasad RK, Agarwal RA (2016) Unregulated emissions and health risk potential from biodiesel (KB5, KB20) and methanol blend (M5) fuelled transportation diesel engines. Renewable Energy 98:283–291

Ajav EA, Singh B, Bhattacharya TK (1998) Performance of a stationary diesel engine using vapourized ethanol as supplementary fuel. Biomass Bioenerg 15:493–502

Baranescu R, Hilger U, Bartunek B, Scheid E, Pischinger F, Rogers G, Last R (1989) Prototype development of a methanol engine for heavy-duty application-performance and emissions. Society of Automotive Engineers Technical Paper No. 891653

Bayraktar H (2008) An experimental study on the performance parameters of an experimental CI engine fueled with diesel-methanol-dodecanol blends. Fuel 87:158–164

Bharadwaz YB, Rao BG, Rao VD, Anusha C (2016) Improvement of biodiesel methanol blends performance in a variable compression ratio engine using response surface methodology. Alexandria Eng J 55:1201–1209

Bozzano G, Manenti F (2016) Efficient methanol synthesis: perspectives, technologies and optimization strategies. Prog Energy Combust Sci 56:71–105

BP (2016) BP statistical review of world energy

Campos-Fernández J, Arnal JM, Gómez J, Dorado MP (2012) A comparison of performance of higher alcohols/diesel fuel blends in a diesel engine. Appl Energy 95:267–275

Canakci M, Sayin C, Ozsezen AN, Turkcan A (2009) Effect of injection pressure on the combustion, performance, and emission characteristics of a diesel engine fueled with methanol-blended diesel fuel. Energy Fuels 23:2908–2920

Centi G, Perathoner S (2009) Opportunities and prospects in the chemical recycling of carbon dioxide to fuels. Catal Today 148:191–205

Chao M, Lin T, Chao H, Chang F, Chen C (2011) Effects of methanol containing additive on emission characteristics from a heavy-duty diesel engine. Sci Total Environ 279:167–179

Cheng CH, Cheung CS, Chan TL, Lee SC, Yao CD, Tsang KS (2008) Comparison of emissions of a direct injection diesel engine operating on biodiesel with emulsified and fumigated methanol. Fuel 87:1870–1879

Cheung CS, Di Y, Huang Z (2008) Experimental investigation of regulated and unregulated emissions from a diesel engine fueled with ultralow-sulfur diesel fuel blended with ethanol and dodecanol. Atmos Environ 42:8843–8851

Ciniviz M, Kose H, Canli E, Solmaz O (2011) An experimental investigation on effects of methanol blended diesel fuels to engine performance and emissions of a diesel engine. Sci Res Essays 6:3189–3199

Dai J, Saayman J, Grace JR, Ellis N (2015) Gasification of woody biomass. Ann Rev

Diesel Progress International (January–February 2016)

Fiedler E, Grossmann G, Kersebohm DB, Weiss G, Witte C (2000) Methanol. Ullmann's encyclopedia of industrial chemistry

Guan G, Kaewpanha M, Hao X, Abudula A (2016) Catalytic steam reforming of biomass tar: prospects and challenges. Renew Sustain Energy Rev 58:450–461

Guerrieri D, Caffrey P, Rao V (1995) Investigation into the vehicle exhaust emissions of high percentage ethanol blends. Society of Automotive Engineers Technical Paper No. 950777

Hansen AC, Zhang Q, Lyne PWL (2005) Ethanol–diesel fuel blends; a review. Biores Technol 96:277–285

Heisey J, Lestz S (1981) Aqueous alcohol fumigation of a single-cylinder DI diesel engine. Society of Automotive Engineers Technical Paper No. 811208

Houser KR, Lestz SS, Dukovich M, Yasbin RE (1980) Methanol fumigation of a light duty automotive diesel engine. Society of Automotive Engineers Technical Paper No. 801379

https://www.mandieselturbo.com/docs/default-source/shopwaredocuments/using-methanol-fuel-in-the-man-b-w-me-lgi-series.pdf?sfvrsn=4

https://ihsmarkit.com/products/chemical-market-daily-service.html

Huang ZH, Lu HB, Jiang DM, Zeng K, Liu B, Zhang JQ, Wang XB (2004) Engine performance and emissions of a compression ignition engine operating on the diesel—methanol blends. Proc Inst Mech Eng, Part D: J Automobile Eng 218:435–447

Jet Propulsion Laboratory and Division of Chemistry and Chemical Engineering, California Institute of Technology Pasadena (1983), California Methanol Assessment Volume II: Technical Report. (https://authors.library.caltech.edu/9320/1/JPLpub83-18vol2.pdf)

Kim S, Dale BE (2005) Environmental aspects of ethanol derived from no-tilled corn grain: nonrenewable energy consumption and greenhouse gas. Biomass Bioenerg 28:475–489

Kisenyi JM, Savage CA, Simmonds AC (1994) The impact of oxygenates on exhaust emissions of six European cars. Society of Automotive Engineers Technical Paper No. 940929

Klier K (1982) Methanol synthesis. In: Advances in catalysis, vol 31. Academic Press, New York, pp 243–313

Kremer FG, Jordim JLF, Maia DM (1996) Effect of alcohol composition on gasoline vehicle emissions. Society of Automotive Engineers Technical Paper No. 962094

Kumar MS, Ramesh A, Nagalingam B (2003) An experimental comparison of methods to use methanol and jatropha oil in a compression ignition engine. Biomass Bioenerg 25:309–318

Levitus S, Antonov JI, Boyer TP, Locarnini RA, Garcia HE, Mishonov AV (2009) Global ocean heat content 1955–2008 in light of recently revealed instrumentation problems. Geophys Res Lett 36:1–5

Lu XC, Huang Z, Zhang WG, Li DG (2004a) The influence of ethanol additives on the performance and combustion characteristics of diesel engines. Combust Sci Technol 176:1309–1329

Lu XC, Yang JG, Zhang WG, Huang Z (2004b) Effect of cetane number improver on heat release rate and emissions of high speed diesel engine fueled with ethanol–diesel blend fuel. Fuel 83:2013–2020

Odaka M, Koike N, Tsukamoto Y, Narusawa K (1989) Optimizing control of NO*x* and smoke emissions from DI engine with EGR and methanol fumigation. Society of Automotive Engineers Technical Paper No. 920468

Prashant GK, Lata DB, Joshi PC (2016) Investigations on the effect of methanol blend on the combustion parameters of dual fuel diesel engine. Appl Therm Eng 103:187–194

Qi DH, Chen H, Geng LM, Bian YZH, Ren XC (2010) Performance and combustion characteristics of biodiesel-diesel-methanol blend fuelled engine. Appl Energy 87:1679–1686

Richards BG (1990) Methanol-fueled Caterpillar 3406 engine experience in on-highway trucks. Society of Automotive Engineers Transactions, pp 1033–1045

Roan V, Betts D, Twining A, Dinh K, Wassink P, Simmons T (2004) An investigation of the feasibility of coal-based methanol for application in transportation fuel cell systems. Report. University of Florida, USA

Saravanan CG, Saravanan B, Sitharthaseelan J, Sudhakar S, Raja A, Sharavanan AR (2002) Fumigation of methanol and fuel additives in a diesel engine testing the performance and emission characteristics. Society of Automotive Engineers Technical Paper No. 2002-01-2722

Sayin C (2010) Engine performance and exhaust gas emissions of methanol and ethanol-diesel blends. Fuel 89:3410–3415

Shahbaz M, Yusup S, Inayat A, Patrick DO, Ammar M (2017) The influence of catalysts in biomass steam gasification and catalytic potential of coal bottom ash in biomass steam gasification: a review. Renew Sustain Energy Rev 73:468–476

Suresh R, Kumar PS, Durgaprsad B, Albert M (2010) Performance analysis for emission and combustion of methanol blend diesel fuel in CI engine using glow plug. In Frontiers in Automobile and Mechanical Engineering (FAME), pp 164–169

Tapley BD, Bettadpur S, Watkins M, Reigber C (2004) The gravity recovery and climate experiment: mission overview and early results. Geophys Res Lett 31:1–4

Taylor AB, Mocan DP, Bell AJ, Hodgson NG, Myburgh IS, Botha JJ (1996) Gasoline/alcohol blends: exhaust emission, performance and Burn-rate in multi-valve production engine. Society of Automotive Engineers Technical Paper No. 961988

Tijm PJA, Waller FJ, Browna DM (2001) Methanol technology developments for the new millennium. Appl Catal A 221:275–282
NOAA/NCEI. https://www.ncdc.noaa.gov/monitoring-references/faq/indicators.php
PMEL. https://www.pmel.noaa.gov/co2/story/Ocean+Acidification
World Meteorological Organization's Greenhouse Gas Bulletin. https://public.wmo.int/en/resources/library/wmo-greenhouse-gas-bulletin
Yao C, Cheung CS, Cheng C, Wang Y, Chan TL, Lee SC (2008) Effect of diesel/methanol compound combustion on diesel engine combustion and emissions. Energy Convers Manag 49:1696–1704
Yasin MH, Yusaf T, Mamat R, Fitri YA (2014) Characterization of a diesel engine operating with a small proportion of methanol as a fuel additive in biodiesel blend. Appl Energy 114:865–873
Zhang ZH, Tsang KS, Cheung CS, Chan TL, Yao CD (2011) Effect of fumigation methanol and ethanol on the gaseous and particulate emissions of a DI diesel engine. Atmos Environ 45:2001–2008
Zhang ZH, Cheung CS, Yao CD (2013) Influence of fumigation methanol on the combustion and particulate emissions of a diesel engine. Fuel 111:442–448
Zhao Z, Lakshminarayanan N, Swartz SL, Arkenberg GB, Felix LG, Slimane RB, Choi CC, Ozkan US (2015) Characterization of olivine-supported nickel silicate as potential catalysts for tar removal from biomass gasification. Appl Catal A 489:42–50
Zhu L, Cheung CS, Zhang WG, Huang Z (2010) Emission's characteristics of a diesel engine operating on biodiesel and biodiesel blended with ethanol and methanol. Sci Total Environ 408:914–921

Chapter 3
Improving Efficiency of Diesel Traction and Adoption of Liquid Sunshine for Indian Railways

Anirudh Gautam, Vagish Kumar Mishra and Avinash Kumar Agarwal

Abstract Indian Railways is facing many challenges. The railway ministry has announced 100% electrification of the Indian Railways traction network by putting an end to the diesel locomotives. This decision needs in-depth debate and discussions as the following paper illustrates. This is necessitated by the fact that India has in-house design and manufacturing capability of diesel locomotives with large export potential. India, like every other country in the world, is unique and has its own set of problems. Solutions to these problems have to be established by taking into account the localised context. Self-powered locomotives in contrast to the catenary-based locomotives offer many advantages and flexibility in operation. The use of methanol as fuel for these locomotives provides unparalleled advantages on all three 'E' dimensions, i.e. economy, efficiency and environment. In this context, this paper discusses the roadmap for efficiency improvement and adoption of methanol as traction fuel for self-powered propulsion for Indian Railways. The paper also compares the two proposed modes of traction for Indian Railways and proves beyond doubt that the elimination of self-powered locomotives from Indian Railways is a faulty decision and must be corrected urgently. IR, on the contrary, should convert its entire fleet of diesel locomotives to operate on methanol, which is a right path to follow.

Keywords Methanol · Locomotives · Diesel traction · Indian Railways

A. Gautam (✉) · V. K. Mishra
Research Designs & Standards Organisation, Ministry of Railways, Lucknow, India
e-mail: ag.srestha@gmail.com

A. K. Agarwal
Engine Research Laboratory, Department of Mechanical Engineering,
Indian Institute of Technology Kanpur, Kanpur 208016, India
e-mail: akag@iitk.ac.in

A. K. Agarwal et al. (eds.), *Methanol and the Alternate Fuel Economy*,
Energy, Environment, and Sustainability,
https://doi.org/10.1007/978-981-13-3287-6_3

3.1 Introduction

The vision of Indian Railways is to march towards ultra-low emission railway traction in a sustainable, cost-effective and reliable manner. Efficiency, economy and very low emission railway traction technologies have to propel drive for national energy security, strategy, reliability and self-sufficiency. To this end, Indian Railways needs to define and execute a sound framework of railway propulsion technologies. Indian Railways have a large fleet of diesel locomotives ($\sim$6000), which are used for catering to passenger and goods traffic. Of late, there was a decision by the Ministry of Railways to electrify all routes of Indian Railways. Subsequently, a decision was followed to stop manufacturing of diesel locomotives at Diesel Locomotive Works, Varanasi, and replace the same with manufacturing of electric locomotives. Similarly, Diesel Modernisation Works at Patiala has also been directed to stop rebuilding of diesel locomotives. Decision to electrify all routes of Indian Railways has been taken based on the assumption that there is surplus electrical power generation in India and also because electric locomotives do not have any visible tailpipe emissions. The rationale suggested was that Indian Railways should switch over to 100% electrified routes so as to achieve economy, efficiency and environmental protection in railway traction sector. Unfortunately, these decisions have been taken without any scientific discussions/ deep scientific/analytical foundation on the subject. This paper attempts to scientifically argue the way forward for railway traction in India.

3.1.1 International Efforts to Reduce Carbon Footprint

Paris-based International Energy Agency (IEA), an autonomous organisation, works in the area of energy security, research and analysis. The agency's aim is to discover pathways, which provide reliable, affordable and clean energy for its member countries (29 countries) and beyond. IEA has published a report (2017), which provides an overview of the energy transition and investments required for a low-carbon energy ecosystem. The report brings out that CO_2 emissions from energy sources are required to be brought down by >60% by 2050, while the GDP of the world will triple during this period. World over, policy makers are searching for effective approaches for decarbonisation of energy production and using renewable energy resources has emerged as a key solution. Renewable energy is affordable and gives us an opportunity to reduce CO_2 emission levels by increasing deployment of renewable and attaining significant economic benefits in an environment-friendly manner. Many countries have stated their Nationally Determined Contributions (NDCs) with higher share of renewable to reduce greenhouse gas (GHG) emissions. India has declared its NDC at Paris accord by reducing the intensity of its GHG emissions by 33–35% on GDP basis by the year

2030 compared to the base year of 2005 (http://www4.unfccc.int/ndcregistry/PublishedDocuments/India%20First/INDIA%20INDC%20TO%20UNFCCC.pdf).

According to the Paris accord on the NDCs, each country needs to support their achievements with data. Accelerated efforts are therefore required to reduce emissions at a quick rate to limit global temperature rise. International Renewable Energy Agency (IRENA) report in 2017 (www.irena.org) pointed out the role of renewable energy technologies and greater energy efficiency in achieving required emission reduction levels by 2030 and 2050. There will be requirement of fossil fuel switching, carbon sequestration, development of efficient energy storage and use of nuclear energy to close the remaining gap. Energy efficiency improvement measures and application of renewable energy will be major factors in reducing emissions and achieving low global carbon energy system. These factors, i.e. energy and materials' efficiency improvements, can lead to reduction of emissions by about 4 Gt by 2030, which translates to ~30% of the emission reduction required (www.irena.org). Electrification of vehicle drives would reduce ~1.5 Gt, which is ~10% of needed cuts. Renewable energy options identified the G20 countries have potential to reduce emissions by additional 10 Gt (www.irena.org).

As a consequence of these interventions, 2030 emissions would reduce by ~25.5 Gt (fossil fuel combustion emitting about ~22 Gt of CO_2 emissions per year). These reductions are sufficient to follow the path of 2 °C global temperature increase prevention by 2030. However to remain on this path, more intense efforts are required between 2030 and 2050. Energy-related CO_2 emissions are required to reduce below 10 Gt by 2050 (70% lower than 2015 levels) and 31 Gt less than the Reference Case. Almost half of these reductions will be through use of renewable energy technologies. Energy efficiency improvements and electrical drive of vehicles would account for the other half. Industrial measures will contribute to the remaining 10% reductions, notably carbon capture and sequestration, material efficiency improvements and structural changes (www.irena.org) (Fig. 3.1).

Although transport sector contribution to total GHG emission is only ~14%, these are only from tailpipe emissions (railways share to the transport sector GHG emissions is <5% of the total transport sector GHG contributions) and do not reflect complete emissions from the entire transport sector. A true picture will emerge only after undertaking a complete life cycle assessment (LCA), which accounts for construction, maintenance and operations of transport sector assets.

3.1.2 Existing Situation of Indian Railways

Indian Railways carries large volume of passenger and freight traffic; however, this share has been declining steadily over the years, leading to railways losing the market share to road transport, thus increasing the burden of harmful emissions from the entire transport sector in India. Modal shift from railways, which is a green

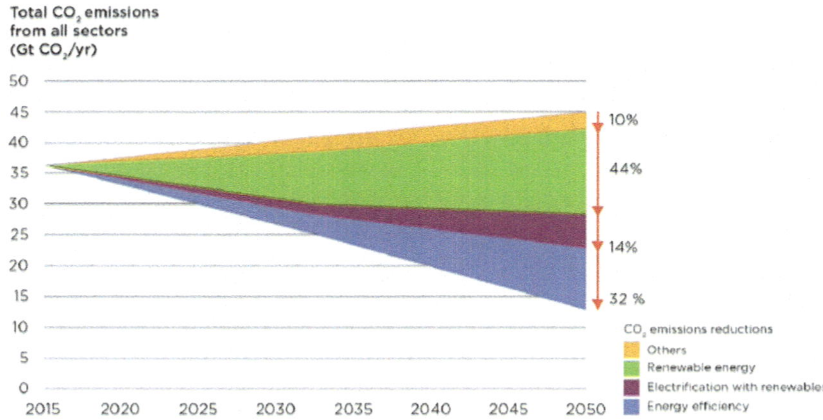

Notes: CO_2 emissions include energy-related emissions (fossil fuel, waste, gas flaring) and process emissions from industry. If only fossil fuel emissions were displayed in this figure, CO_2 emissions would start from 33 Gt in 2015 and would reach 40.5 Gt and 9.5 Gt per year in 2050 in the Reference Case and REmap, respectively.

Fig. 3.1 Primary CO_2 emission reduction potential by technology in the Reference Case and renewable energy map (2015–2050) (www.irena.org)

mode of transport to the roadways, which is carbon intensive transport mode, is depicted in Figs. 3.2 and 3.3.

This modal shift is due to several reasons, which include flexibility of transportation by road, inefficient logistics management by railways, significant increase in budgetary support to roadways and a large reduction in budgetary support to railways. A modal shift in favour of road transport has led to increase in GHG emissions, as well as traffic congestions, and environmental degradation due to worsening living conditions caused by toxic pollutants emitted by heavy-duty and

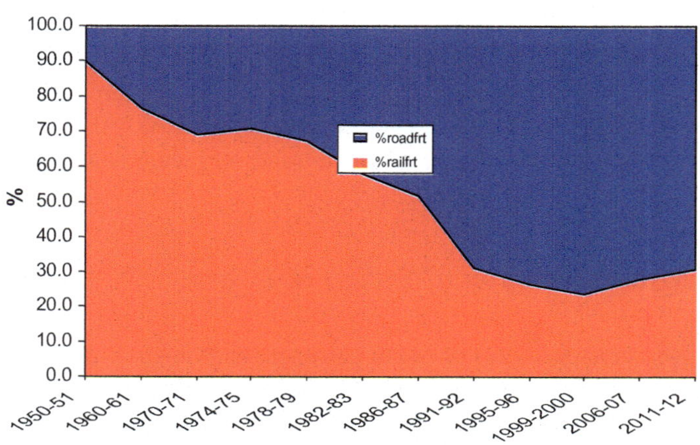

Fig. 3.2 Percentage share of rail and road freight traffic in India over last six decades

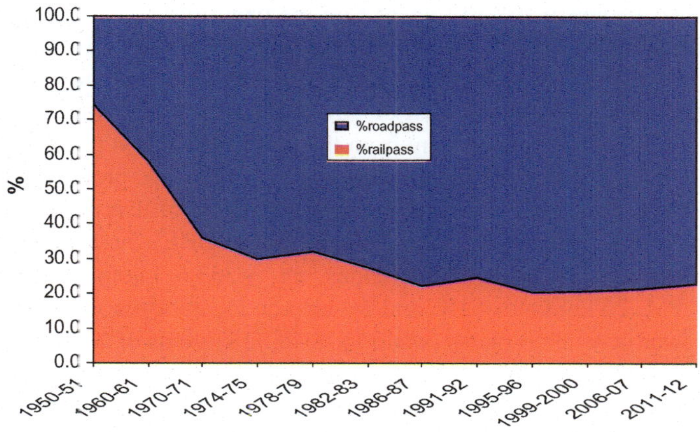

Fig. 3.3 Percentage share of rail and road passenger traffic in India over last six decades

medium-duty road transport vehicles. In spite of modal shift in favour of road transport, Indian Railways still carries ~8 billion passenger and 1 billion-ton freight annually. This is made possible by a fleet of 5500 electric locomotives and 5200 diesel-electric locomotives. Unlike road vehicles, both types of locomotives on Indian Railways are electrified. The electrical locomotive draws current from the catenary through a transformer to power the traction motors and the wheels, while the diesel-electric locomotive has an on-board diesel engine, which generates electricity to power the traction motors and the wheels. The efficiency of transmission on both locomotives is >90%, therefore the efficiencies of these power-plants are significantly higher than the corresponding road vehicles. Diesel-electric locomotives utilise large-bore mid-speed diesel engines, which have peak efficiencies of ~40 to 42%, compared to road vehicle engines where efficiencies are typically in the range of ~25 to 30% and the gas turbines efficiencies in the range of ~30 to 33%. Modal shift to road sector increases inefficient use of scarce and imported petroleum resources. Therefore as a first step, it is essential to increase the model share of passenger and freight traffic in Indian Railways.

Of the two traction modes used in the Indian Railways, diesel traction is being derided as a polluting and expensive mode. However, these conclusions are based on notions and devoid of any serious scientific analysis. An attempt has been made to calculate the tank-to-wheel energy consumption and expenditure related to the two types of locomotives. Life cycle assessment of different locomotive fuels and technologies has also been initiated in Indian Railways, and the report shall be presented sometime in future. Diesel locomotives consume ~2.8 billion litres of diesel per year with an annual expenditure of ~Rs. 16,000 Crores. The expense of Rs. 16,000 Crores includes ~100% tax imposed by the Government of India on petroleum products. Corresponding annual fuel consumption of electric locomotives of IR comes to ~5.8 billion litres of diesel equivalent of electricity, and the yearly electricity bill of Rs. 10,500 Crores, fixed structures annual maintenance cost of ~Rs. 1000 Crores and yearly interest on the fixed infrastructure to the tune of ~Rs. 20,000 Crores or more (total ~Rs. 31,500 Crores) are incurred. The fixed infrastructure cost includes the cost of setting up structures and catenary on Indian Railways network, proportional cost of power generating plants and proportional cost of electricity distribution system from the power generation plants (~Rs. 1.25 Crore per km of Over Head Equipment, ~Rs. 3.57 Crore proportional cost of power plant per km, ~Rs. 3.36 Crore proportional cost of distribution infrastructure per km of OHE: Total ~Rs. 8.18 Crore per km of OHE).On the emissions side, diesel locomotive tank-to-wheel emissions are ~353.6 kilotons of NO_x, 15.9 kilotons of PM, 25 kilotons of HC and the electric locomotives share of emissions from the power generation plants comes to ~483.2 kilotons of NO_x, 80.5 kilotons of PM, 483.2 kilotons of SO_x and 24 tons of Hg (pollutants emitted for setting up overhead structures and their maintenance have not been taken into account). Pollutants from the electric traction have been calculated by assuming that the power plants meet the limits set by the central pollution control board.

Electricity generation mix of 60% coal, 10% oil and gas and 30% hydro, nuclear and renewable has been taken, based on the reports published by the Central Electricity Authority and Ministry of Power. Actual generation of power from renewable is ~6.9% (wind-60%, solar-14.6%, biomass-4.8%, bagasse-9%, small hydro-10.7%, others-0.44%), hydro 11.1% and nuclear 3% (total ~20%); therefore, the balance 80% electricity generation is from coal primarily (~75%) and gas (~5%). This is against an installed capacity of renewable of hydro 14% and nuclear 2% (Government of India, Power Sector 2017). Renewable and hydro being intermittent sources of energy are not able to operate at high load factors, and this is the main reason for lower actual electricity generation compared to the installed capacity of electricity generation (www.eia.gov) (Fig. 3.4).

In a study done by Das and Roy (Das and Roy 2018), it has been estimated that electricity generation will continue to be predominantly coal based, ~50%, even

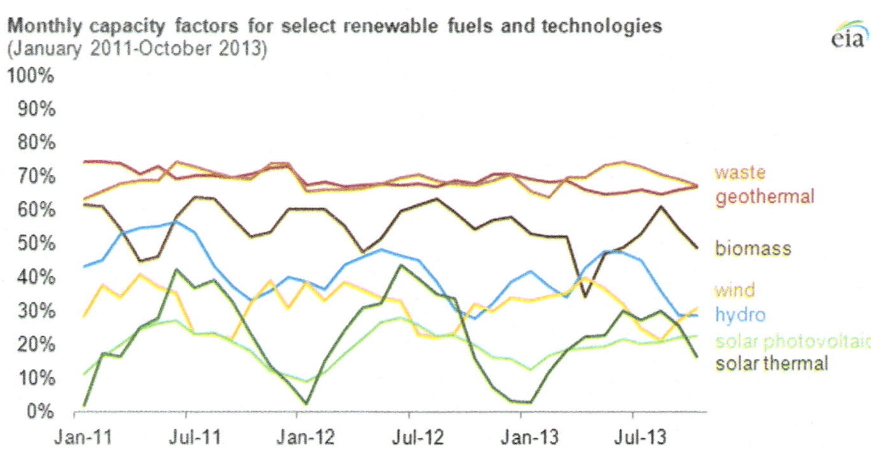

Fig. 3.4 Monthly capacity factors for select renewable fuels/technologies (www.eia.gov)

up to year 2050 albeit with super critical and CCS technologies. This implies that even with 100% electrification, Indian Railways will not be able to meet its NDC, as committed in the Paris accord. Additionally, it will not be possible to meet Indian Railway's vision of ultra-low emissions from traction, achieving economy and efficiency in its propulsion systems. Therefore, there is a need to carry out research for increasing the efficiency and fuel switch for the existing propulsion systems as well as developing alternate propulsion technologies, which can meet the NDC commitment and cater to the vision of Indian Railways for propulsion systems. Switching over to a single mode of traction, i.e. electrical traction, definitely weakens the national security, strategy and reliability of propulsion systems used by Indian Railways. Aspects such as grid and infrastructure failures/sabotage, troop movement to border areas in case of war and disaster management in times of national calamities also need to be taken into account. A healthy mix of traction technologies for Indian Railways will help obviate these issues.

3.2 Future of Self-powered Traction for Indian Railways

Technavio, a leading consultancy firm, has brought out a report entitled 'Global Diesel Locomotive Engines Market 2016–2020' on growth of diesel locomotive industry (https://www.technavio.com/report/global-powertrain-global-diesel-locomotive-engines-market-2016-2020) worldwide. This industry research report identifies General Electric, Caterpillar and China CNR as the top suppliers in global diesel locomotive market. The report did a competitive analysis of the market on two dimensions, i.e. application (passenger and freight) and geography (the Americas, APAC and EMEA). The report predicted that the global diesel locomotive market will grow at a steady rate during the next four years and will post a CAGR of $\sim$3% by 2020. High cost of overhead rail electrification was identified as one of the primary growth factors for this market. The report gave an example of electrification of 'Great Western mainline' in Britain, whose budget exceeded almost three times the estimated value to $\sim$US$4 billion. Additionally, it also raised environmental concerns regarding levelling the natural abode for construction of electric overhead lines. Such factors are likely to increase the demand for diesel locomotives, propelling its market growth.

Development of hydrogen-powered locomotives is likely to emerge as a future traction technology in coming years (Future Power Technologies 2013). The hydrogen stored in the roof will generate electricity by chemically reacting hydrogen and oxygen in a fuel cell to generate electricity and will emit only condensed water and steam as by-products. This also eliminates the need for overhead cables, making these trains environment-friendly. Locomotive manufacturer 'Alstom' has taken the lead by launching 'Coradia iLint', a hydrogen fuel cell-powered passenger train in Germany in 2016. Maximum speed of the train is $\sim$145 mph with a range of $\sim$500 miles prior to refuelling.

The diesel locomotive market has a limited number of well-established vendors and is dominated by major players namely General Electric (GE), Caterpillar and China CNR. Indian Railways with two diesel locomotive manufacturing units at Varanasi and Patiala does not figure in the list of prominent locomotive manufacturers and suppliers. On the other hand, China CNR has been steadily increasing its market share in the global diesel locomotive market. Due to restricted demand for diesel locomotives, this market is moderately competitive. Major diesel locomotive manufacturers are presently focusing on increasing their customer base and investing heavily in R&D. Due to technological advancements and regulations adapted by various governments, the market's competitive environment will become more intense in the near future. A sizeable number of supplier base industries have been established around diesel locomotive manufacturing-based economic activities in India. In this context, Ministry of Railways' decision to close down the diesel locomotive manufacturing harms the economic interests and "Make in India" programme of the country, while offering the global diesel locomotive market to Chinese and other players on a platter and at the same time, purging the entire ecosystem built around diesel locomotive industry in India. Many manufacturers in diesel locomotive industry are also in the export market or have potential to become exporters, but this decision will lead to them shutting their business.

3.3 Improving Efficiency of Diesel Locomotives of Indian Railways

As per the global blueprint given by IEA, $\sim$32% of INDC (IEA 2017) goals are to be met by improving the efficiency of existing/future motive power plants. Existing diesel locomotives and diesel-electric multiple units (DEMUs) of Indian Railways are required to be updated to latest engine technologies that are commercially available. There is an immediate requirement of research in developing state-of-the-art engine technologies as well as creation of world-class research facilities for upgrading the self-powered propulsion technologies for Indian Railways.

3.3.1 Diesel Locomotive Engines

Diesel locomotive fleet on Indian Railways consists of following types of locomotives: (a) 3300 hp rated mainline locomotives, (b) 4500 hp mainline locomotives and (c) 1350 hp shunting locomotives. These locomotives are powered by large-bore medium-speed locomotive engines with high fuel efficiencies and low exhaust emissions. However, the technology of these locomotive engines is quite old and therefore is far from "best in class". There is large scope to increase the horsepower/torque of these engines, reduce fuel consumption, reduce harmful emissions and increase the reliability. The fleet of these diesel locomotives will remain with Indian Railways for next thirty years and will continue to consume petroleum-based fuels and increase the CO_2 burden of railways.

With a view to meet the objectives of reducing fuel consumption and emissions and to increase the reliability of locomotives, a state-of-the-art engine laboratory and engine development directorate was established in RDSO in 1987. This directorate was responsible for uprating of the horsepower of the imported locomotive engines, improving their technology for reduction in fuel consumption, enhancing reliability and indigenising these engines. Consequence to the R&D carried out by the Engine Development Directorate of RDSO, the rated power of the locomotive engines was increased to 3300 hp from base of 2400 hp for one type of locomotive engine and to 4500 hp for the second type of locomotive engine. The brake-specific fuel consumption (BSFC) of these engines was also reduced by ~ 10 to 15%. The fuel savings and increased rated power led to higher rate of increase in goods and passenger traffic and corresponding lower rate of increase in fuel consumption for diesel traction by Indian Railways. However, these locomotive engines were still below the global benchmark in terms of fuel economy, emissions and reliability. Feasibility studies have been conducted by RDSO to upgrade these engines to global standards. The findings from these studies indicate that IR's locomotives can be further upgraded with minimal investment, leading to substantial improvement in the above performance parameters. On the other hand, premature condemnation of these locomotives/re-engaging/inefficient use shall result in significant loss to Indian Railways and Government of India. Hence, the following technology and research areas are required to be taken up for development and upgradation of existing diesel locomotives to meet global objectives of fuel economy, GHG emissions and reliability.

3.3.2 *Measures to Improve Efficiency and Reduce Emissions*

3.3.2.1 Short-Term Targets

There is an immediate need to incorporate latest engine technologies in locomotive engines with a targeted reduction of 10% diesel consumption and 50% emissions and targeted increase of 10% in rated power output of locomotive engines. Both types of locomotive engines are equipped with mechanical fuel injection system; therefore, there is a need to provide flexible fuel injection system such as 'common rail electronic fuel injection system or electronic fuel injection system' on all locomotives along with high-efficiency turbochargers. This would not only reduce fuel consumption and emissions but also increase reliability of these engines.

Engine Development Directorate of RDSO has tested a prototype common rail fuel injection system and variable geometry turbocharger on ALCO locomotive engine. These showed a reduction in brake-specific fuel consumption and particulate matter emissions. Retrofitment of variable geometry turbocharger on all ALCO locomotives along with common rail fuel injection system will increase fuel savings and reduce harmful emissions substantially. Introduction of electronic fuel injection system along with variable geometry turbocharger requires high effectiveness intercooler and optimisation of combustion chamber geometry and fuel spray. At present, the valve timings of the ALCO and EMD locomotive engines are optimised for the rated load (i.e. eighth engine notch). This adversely affects the volumetric efficiency of the engine at lower engine notches; hence, fuel efficiency and emissions are adversely affected. In the large-bore engines, there is a trend to incorporate variable valve timing system, which can be optimised at all engine notches. There is a need to introduce variable valve timings on both types of diesel locomotives of Indian Railways. Introduction of these measures is expected to reduce the fleet fuel consumption of Indian Railways by $\sim$10%, leading to monetary benefits of $\sim$Rs. 2000 Crores (US$300 million) per year. Another important area is the weight reduction of the body structure of the locomotive. The engine hood, radiator compartment and drivers cab are made up of mild steel sheets. Lightweight composite materials such as reinforced carbon fibre can be used in place of steel. Other composite materials can also be considered. The duty cycle of the diesel locomotives includes frequent change in the engine notches and long periods of idling (>50%). Even during idling, the locomotive fuel consumption varies between 20 and 30 L/h. With a view to reduce fuel consumption, Indian Railways has started installing auxiliary power unit (APU) on the diesel locomotives. APU consists of a small piston engine, a generator, air compressor and a control unit. During engine idling, the locomotive main engine shuts off and the APU engine is automatically started. The APU charges locomotive batteries and the air reservoir for train brakes. APU saves $\sim$10% fuel consumption and reduces the associated harmful emissions from diesel locomotives. One of the ways of reducing fuel consumption is to optimise the operation of diesel locomotives so that unnecessary accelerations and braking can be avoided. Indian Railways has developed a system for optimised driving for the loco pilots (GOLD), which has demonstrated reduction in fuel consumption by as high as $\sim$15% over the duty cycle. Indian Railways has fitted ‘GOLD’ on many locomotives; however, it makes economic sense to fit these on all diesel locomotives of IR. Condition monitoring of locomotives assists in identifying faults in advance of actual failures, thus preventing excessive fuel consumption and emissions during the interim period of failures. A system called ‘Remote Monitoring of Locomotives’ (REMMLOT) with the capability for remote condition monitoring of diesel locomotives has been developed by RDSO in association with industry. In the first phase, all diesel locomotives must be fitted with REMMLOT.

One of the main complaints regarding use of diesel locomotives is the noise emanating from its diesel engine. Modern diesel engines are as quiet as gasoline engines; thus, a project has been taken up to reduce the noise, vibrations and harshness of these diesel locomotives jointly with IIT Kharagpur. This will reduce the noise pollution from these locomotives and make them more acceptable for use in urban areas.

Another major problems faced is high-quality maintenance of railway assets and procurement of spare parts for locomotives. Government of India and Ministry of Railways continue to use archaic procurement procedures, which are time-consuming, have very high cycle times and often have potential for financial corruption. World over, private as well as government sectors have streamlined their procurement processes. Automotive Industries uses the ISO standard (ISO 16949) for procurement of materials. Electro-Motive Diesels, General Electric and other prominent locomotive manufacturers use 16-step vendor development process. These processes are objective and ensure that only quality suppliers are able to qualify for the supply chain and leave no room for manipulation and corruption. At the same time, these modern procurement processes employ systems for advanced procurement planning and quality control, in order to ensure high quality of the supplied products. The lifecycle cost of high-quality products is lesser than other options available in the market.

3.3.2.2 Medium-Term Targets

Hybridisation of Diesel Locomotives

At Daimler AG, AnkeKleinschmit, Vice President, Group Research and Sustainability and Chief Environmental Officer (https://www.sae.org/news/2016/08/sustainable-transformationthe-daimler-wy) stated "*that in long run, electrification of the drive-train is one major element for the 'future of mobility*" (https://www.sae.org/news/2016/08/sustainable-transformationthe-daimler-wy). Daimler is of the view that different requirements will need different technology paths; therefore, Daimler is pursuing a flexible strategy to reach the goal of zero-local-emissions driving. For this, Daimler is (1) improving the existing high-tech combustion engines, (2) increasing hybridisation to further reduce fuel consumption and (3) developing technology for zero-local-emissions driving with batteries and/or fuel cells. Daimler predicts that it will be some more time for 100% electric vehicles to have high market share; however with growing range of EVs, the proportions are continuously improving (https://www.sae.org/news/2016/08/sustainable-transformationthe-daimler-wy). Companies such as Volvo, Cummins and Caterpillar also echo similar views on electrification of vehicles. Hence, it is essential that Indian Railways also considers hybridisation of its diesel locomotives to reduce fuel consumption and their carbon footprint. This will require on-board energy storage, which can be chemical or hydro or mixed forms. Locomotive manufacturers, like GE, ALSTOM, East Japan Railway, RailPower Corp. USA, Hitachi high-speed locomotive, have already developed prototypes of hybrid locomotives/rail cars, and this trend is expected to continue further. Hybridisation

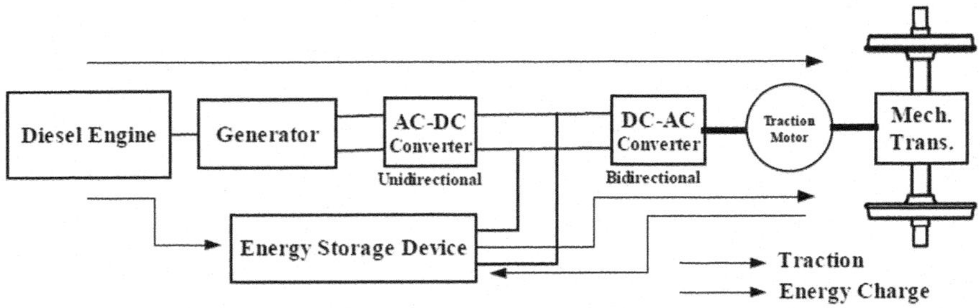

Fig. 3.5 Schematic of hybridisation concept of self-powered locomotives (Future Power Technologies 2013)

allows introduction of regenerative braking on the self-powered locomotives with energy savings to the extent of ~30 to 40%. This in turn helps in fuel savings as well as reduction in GHG emissions. Concept of hybridisation of self-powered locomotives is given in Fig. 3.5.

In the above concept, diesel engine can be converted to run on methanol/DME, and for this, some modifications in the engine hardware are essential. In cases of new self-powered locomotives, heavy-haul and high-speed hybrid gas turbine locomotives can be designed and developed. The modes of operation of hybrid locomotives are as follows: (a) battery mode alone-motive power is off, and locomotive is powered by battery only, (b) engine alone mode-power is drawn from engine/gas turbine, (c) combined mode-power from both engine/gas turbine and the energy storage device provides power to the traction motor, (d) power split mode-engine/gas turbine power splits to drive the train and charge the battery, (e) stationary charging mode, and (f) regenerative braking mode-braking energy is stored in the on-board energy storage devices. The picture of a hybrid diesel locomotive developed by General Electric Transportation Company (USA) is shown in Fig. 3.6.

Use of Cleaner/Renewable/Alternate Fuels

Hydrogen is the ultimate clean fuel with zero tailpipe emissions. Hydrogen can be used as a fuel in IC engines as well as fuel cells, and it does not leave any carbon footprint. However, there are issues with production and storage of hydrogen. Abbott suggested novel ways of production of hydrogen from solar power and termed it as 'solar hydrogen' (Abbott 2010). Concentrated solar power is used to run Stirling engines, which produce electricity, that can be grid connected or else can be used for electrolysis of water to produce 'solar hydrogen'. To overcome the problem of storage and carrying of hydrogen, renewable hydrogen is combined with atmospheric carbon dioxide to produce methanol. Methanol/DME is being considered as the next-generation electro-fuels (Todd Davidson et al. 2017). Both methanol/DME does not contain any carbon-to-carbon bonds and therefore do not produce any smoke and emit very low concentration of particulates. Methanol is already being used in commercial fuel cells with on-board reformer to convert it

Fig. 3.6 Hybrid diesel-electric locomotive developed by GE Transportation, USA (http://www.railwaygazette.com/news/single-view/view/ge-unveils-hybrid-locomotive.html)

into hydrogen. Biodiesel is another renewable fuel, which significantly reduces the carbon footprint. Indian Railways has mandated use of 5% biodiesel blends on its diesel locomotives. This target can be safely increased to 20% biodiesel blends, if it backed by in-house R&D programme in the Engine Development Directorate of RDSO. The use of methanol on diesel locomotives is likely to reduce the diesel expenditure by ~50%, NO_x by ~70%, particulate matter by ~95% and HC by ~90%. This will be an excellent move towards ultra-low emissions from Railway traction.

Re-engining with Multi-genset Concept for Older Locomotives

Indian Railways has developed a preliminary design for low-power multi-genset locomotives. In this concept, two or three small high-speed engines could be used to produce power, which is combined using a DC link voltage bus. Depending on the power requirement, either a single engine or all the engines are operated, leading to great saving on fuel and emissions. Coupled with on-board energy storage devices, a further reduction in fuel consumption and emissions can be obtained by introducing regenerative braking. Motive Power Directorate and Engine Development Directorate of RDSO have developed some expertise in this area, which needs to be expanded further for developing in-house capabilities for the prototyping of mainline multi-genset locomotive for Indian Railways.

Weight reduction of Locomotives

Fuel consumption of the locomotive is dependent on its weight. Hence, there is need to reduce the weight of locomotives without sacrificing its adhesion power. The use of lighter materials such as aluminium, reinforced carbon fibre, composites could be considered for reducing the locomotive weight. This is also important, if on-board batteries could be employed for hybridisation and regenerative braking.

Reducing Aerodynamic Design of Locomotive

Indian Railways has planned to increase the train speeds to 200 kmph on semi-high-speed routes. As the speed of the train increases, locomotives of good aerodynamic shapes have to be developed in order to reduce drag-related friction and consequent reduction in fuel consumption. RDSO has taken up a collaborative project with IIT Kharagpur to develop aerodynamic profile of the rolling stock, capable of attaining peak speeds of 200 kmph.

Use of AC–AC Traction

AC–AC traction is more efficient than AC–DC traction. At present, IR fleet of diesel locomotives consists of roughly half AC–DC traction locomotives. The use of AC–AC traction increases the adhesion control, reduction in maintenance and higher reliability; therefore, it reduces fuel consumption and emissions. A development programme is required to convert all AC–DC locomotives to AC–AC type. Power electronics of the AC–AC systems has also advanced over the years; hence, there is need to carry out further R&D on for adaptation of these systems, which are superior than IGBT-based systems.

Use of Permanent Magnet Motor

At present, Indian Railways use asynchronous motors for traction. Variable voltage and variable frequency controls are used to convert the torque–speed characteristics of the traction motors in order to suit the traction characteristics. Developments in the wheel-mounted permanent magnet asynchronous motors with higher efficiency enabled us an alternative to conventional AC asynchronous motors with gear transmission, consisting of flexible couplings and reduction gears. Wheel-mounted construction does not require use of transmission system since the torque is directly transmitted to the wheel. VVVF control of these motors allows the use of rare earth permanent magnet motors. This results in design and manufacture of light and efficient (less mass and volume) traction motors. As a result of direct coupling to the wheel, the permanent magnet motors are able to reduce transmission losses, energy consumption, noise and maintenance.

Unified Control of Locomotive Systems

There is a need to transition from a decentralised architecture following the general paradigm "one ECU for one Advanced Driver Assisted System-related

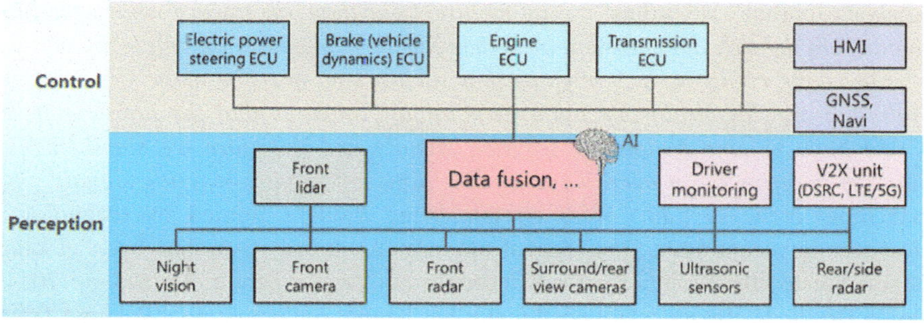

Fig. 3.7 Unified control scheme of a self-powered locomotive (https://www.automotive-iq.com/tag/adas)

functionality" to a centralised architecture with central sensor data fusion, predictive 360° redundant sensing and artificial intelligence (AI), as shown in Fig. 3.7.

Unified control of the self-powered locomotives would include the 'Fog Vision' using millimetre waves for visualisation of the railway path during fog/heavy fog. This will also help in increasing overall efficiency of the locomotive and reduce fuel consumption and emissions.

3.3.2.3 Long-Term Targets

Multiple Modes of Railway Traction

It is envisioned that in the long-term, there will be fusion of self-powered and catenary-based locomotives. Battery technology will require some time to further evolve and increase its power density. The energy density of the batteries and cost of electrification would continue to increase, leading to hybrid locomotives emanating on the scene. These hybrid concepts may range from piston engines and on-board energy storage hybrids, gas turbines and on-board energy storage hybrids, fuel cells and on-board energy storage and hybrids of catenary-based and piston engine/turbine-based locomotives. This will become essential to meet 2050 target of reduction in atmospheric GHG emissions, and efficiencies. In the long-term, it is recommended that at least 2–3 modes of traction be used with a healthy mix of these modes as a strategy for country's security, higher reliability and cater to natural calamities. Multiple modes of traction will ensure safety against failure of single mode of traction for some reason.

Autonomous Control of Locomotives

The railway industry is primed for a crewless train. The idea of an autonomous train has been prevalent since the 1960s. In 1962, the assistant to the president at Westinghouse wrote an article addressing self-driving trains in the 'Financial

Analysts Journal'. He writes: "... *let us direct our thoughts toward what would be required in the way of equipment or system for the safe operation of a train with no person at the controls. It is obvious that such a step must be made if we are to approach crewless train operation. It is equally obvious that the crewless train would have no one to look at wayside signals and therefore we should begin thinking in terms of an overall system that does not require wayside signals*". At that time, crewless trains were an interesting thought limited by technology. However in recent years, the idea of autonomous locomotives has been gaining steam due to the evolution of advanced technology capabilities. In June 2011, Union Pacific Railroads announced that it has begun implementing a Positive Train Control (PTC) line between Spokane, WA and East Point, ID in USA. Unmanned long-haul freight is becoming more than just a pipe dream. In Western Australia, the Rio Tinto line has been hauling iron ore on crewless trains using its patented 'Auto-haul Technology'. Following this development, domestic train companies are looking more closely at the feasibility of such technology. Indian Railways with vast expertise of Indian information technology can certainly develop autonomous technologies for locomotives. This should be an active area of research for Indian Railways.

3.4 Liquid Sunshine for Indian Railways

Currently, ~70% of India's petroleum consumed each year fuels >90% of the country's transport sector in the form of final products such as jet fuel, diesel, gasoline and other products. However, growing concerns over geopolitical uncertainties and climate change with continued petroleum usage have presented fresh challenges for the policy makers, industry stakeholders and consumers. Questions regarding the sustained use of petroleum have re-ignited interest in alternative fuels and explorations into the non-conventional fuel resources. There has been a proliferation of options but little consensus on a path forward. Shifting to alternative fuels can address energy security and climate change issues, but it entirely depends on how they are produced, distributed and used. Alternative fuels and vehicles have to satisfy two broad criteria: (i) to demonstrate an ability or potential to improve energy security or GHG emissions and (ii) to become commercially viable. Table 3.1 illustrates different fuels being used in India in transport sector.

From an emission's standpoint, electric vehicles generally have zero tailpipe emissions, but WTW emissions depend strongly on the local electrical energy supply mix. In India, where 80% electricity is generated by thermal power stations, the WTW emissions of electricity operated vehicles are inferior to those of e.g. hybrid, flex-fuel and alternate fuel vehicles. Methanol offers a clean, fuel-efficient and indigenous way to fuel the IR fleet of locomotives.

Table 3.1 Different fuels being used in India in the transport sector

	Chemical composition	Fuel source (Average feedstock 2011)
Gasoline/diesel	Liquid hydrocarbon mix C_4 to C_{12}	100% conventional crude oil
XTLs (hydrocarbon liquids)	Liquid hydrocarbon mix C_4 to C_{12}	Coal (CTL), natural gas (GTL)
Ethanol	Liquid alcohol CH_3CH_2OH	100% from molasses (sugarcane)
E5/E10	Blends of ethanol and gasoline	January 2003, GOI mandated blending of 5% ethanol in gasoline in nine states and four union territories
CNG	Nearly all methane (CH_4)	60% conventional and tight natural gas, 40% imported as LNG, mainly from Qatar
Electricity	Elementary charged particles generated by friction, induction or chemical change	70% coal; 8.2% gas; 0.3% oil TOTAL 69.4% thermal; 11.9% hydro; 1.9% nuclear; 7% renewable
Hydrogen fuel cell (rarely used, mainly for experiments)		50% natural gas; 30% liquid hydrocarbons; 18% coal

3.4.1 Methanol-Liquid Sunshine Fuel

In the process of photosynthesis, CO_2 combines with water in the presence of solar energy from Sun to produce glucose, a stable form of chemical energy. Glucose is transported through the plant's circulatory system, to individual cells, where it is converted back into energy for powering/regenerating cells and for other vital functions of the plants. In the processes, CO_2 and water are recycled back to the environment. This cycle has continued since the appearance of life on planet earth. Liquid Sunshine vision is about ecologically balanced (carbon neutral) green energy system encompassing energy harvesting, conversion, storage, distribution and utilisation that is compatible with mother nature's cycle (Shih 2017). The ecological balance is achieved by using CO_2 and water to bind and store sun's energy as green liquid fuel. In such a system, high-energy-density stable liquids are stored in energy reservoirs and drawn and distributed on demand.

Methanol is a naturally the first target of Liquid Sunshine. It emits less CO_2 than coal and oil upon combustion per unit energy produced with practically no emissions of NO_x, SO_x, VOCs and PM. It is also a versatile multi-purpose energy carrier for various applications such as (i) feedstock for petrochemicals, (ii) fuel for stationary heat, power and machinery, (iii) fuel for marine and ground transportation, (iv) potential medium for large-scale energy storage as "liquid electricity" and "liquid hydrogen" carrier. The simplest of alcohols, methanol, is a building block for higher-energy-density alcohols, such as ethanol, propanol and butanol (Shih 2017).

Coal, oil and gas are non-renewable and are rapidly diminishing fossil fuels. In the business as usual scenario, GHG emissions will increase exponentially (Fig. 3.8). Adoption of methanol as an energy source provides an opportunity to reduce GHG emissions to meet the 2° temperature target for global survival (Fig. 3.8).

To realise the advantages of methanol as clean fuel, a roadmap has been presented by Shih at Stanford University last year (Shih 2017) (Fig. 3.9). This scheme illustrates the economics and environmental impact (3E) of energy harvesting, storage, distribution and utilisation in a Methanol economy scenario.

In the production of green methanol, CO_2 and H_2O are used to bind and carry energy from the Sun to the end-user. Methanol is also a building block for higher-energy-density alcohols such as ethanol, propanol and butanol. Initially fossil fuels like coal and natural gas are used to produce methanol and later, and renewable energy and atmospheric and industrial CO_2 are utilised to produce green methanol.

The Liquid Sunshine roadmap by NITI Ayog, Government of India, outlines actionable step-wise strategy to (1) deploy commercially viable low-emission technologies to produce clean methanol from natural gas and renewable power as early as 2020 and (2) target the first deployment of scalable green technologies to produce green methanol from sunshine and wind energy by 2040. Green methanol can serve as the energy reservoir for green electricity, forming a synergistic dual energy system-green electricity and green liquids-that fulfil all energy needs for modern-day applications.

Clean methanol can be produced across India. Existing logistics, infrastructure for storage, shipping and distribution can be leveraged. Green technologies and

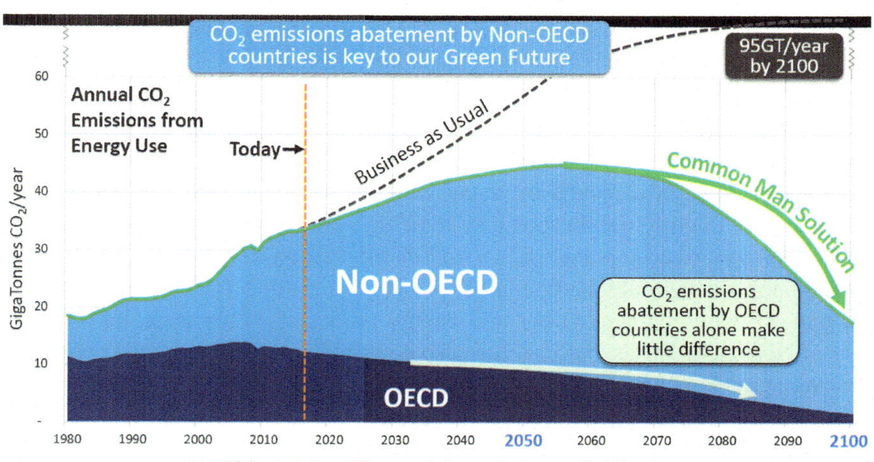

Fig. 3.8 CO_2 emissions from OECD and non-OECD countries (Shih 2017)

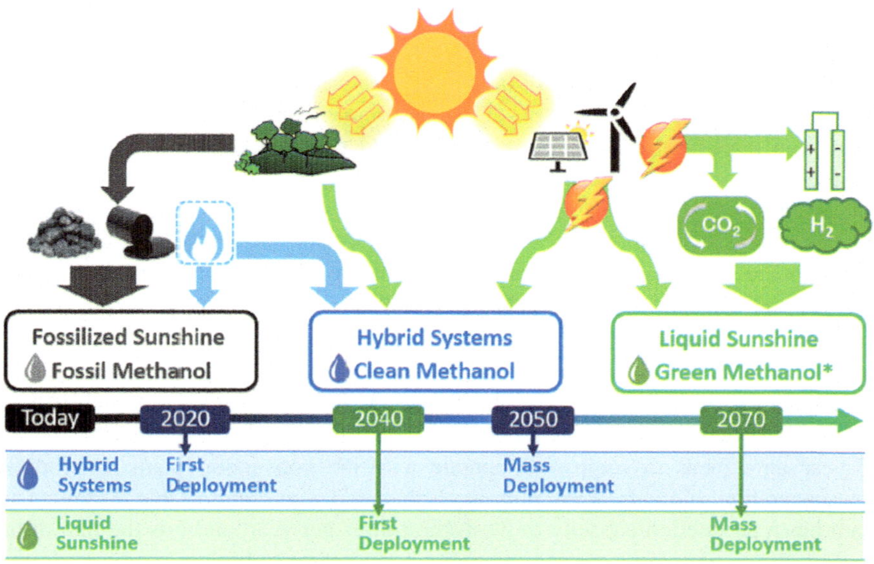

Fig. 3.9 Technology pathways for green methanol: the first target of Liquid Sunshine (Shih 2017)

pathways to produce green alcohols can be adopted, offering new energy avenues for powering the Industrial growth of India. Indian Railways can convert its existing fleet of diesel locomotives to operate on methanol in a straightforward manner. This will require modifications in the existing fuel injection system and increasing the storage capacity of the on-board fuel tank. Both modifications can be easily accomplished by DLW Varanasi and DLMW Patiala.

3.4.2 Use of MSW to Produce Renewable Methanol for Traction

Urban India produces ~50 million tonnes of municipal solid waste (MSW) annually. It grows with economic development at 3–5%. Methanol production pathways from MSW are similar to that produced from biomass (Chen 1995). After removing the inorganic material, MSW is very similar in composition to biomass. For both feedstocks, gasifier is the only non-commercial component in the fuel production system. Different types of gasifiers have been developed for producing methanol from MSW. The thermodynamics of producing methanol from MSW has been analysed using chemical process simulation software and pinch analysis. The thermal efficiency of methanol production process is ~50%. These efficiencies are only slightly lower than the efficiencies that have been calculated elsewhere for biomass feeds. Emission measurements of the gasifiers have shown that they emit fewer gaseous pollutants than modern incinerators and produce liquid and solid wastes that are non-hazardous. Even if 20 million tonnes of methanol can be produced from MSW, this will be sufficient to provide fuel for the self-powered locomotives/train sets of the IR ($\sim$3 million tonnes), and the extra methanol produced can then be used in a cleaner manner to generate electrical power in gas turbine power plants. Methanol produced from MSW has the lower GHG footprint compared to diesel, gasoline, electricity and other fuels derived from conventional sources.

There are many commercial technologies to produce methanol from MSW. Enerkem of Canada converts 30% of MSW generated by city of Edmonton, Canada, to methanol using a fluidised bed gasification process. Gasification produces syngas, which is then converted to methanol. Another way to convert MSW to methanol is by incineration. Incineration of MSW produces CO_2, which is then converted to syngas and then to methanol. Enerke and the City of Edmonton have a 25-year agreement to setup and operate a MSW conversion plant to produce next-generation biofuels from non-recyclable and non-compostable MSW. The plant has a production capacity of 38 million litres per year, and it is the first of its kind of efforts between a metropolitan city and a waste-to-biofuel producer. In the beginning, the plant produced methanol from MSW, and later, it started production of ethanol as well. It is therefore recommended that Indian Railways collaborates with academia and industry to set up methanol production plants from the MSW produced at railway stations and nearby municipalities. This will obviate the twin problems of environmental degradation and MSW management.

3.4.3 Methanol Production from Coal

Higher price of natural gas and oil has evinced renewed interest in alternative feedstocks for production of methanol. Although different types of biomass are being considered to produce methanol, in short-term, coal provides the only viable alternative to produce methanol on large-scale economically. China has been producing methanol from coal for many years, thus proving the feasibility of this technology (Fig. 3.10).

In the gasification process, coal reacts with oxygen and steam in a gasification reactor to generate syngas, which contains H_2, CO and CO_2 and small amounts of inert gases such as nitrogen, argon and methane. Methanol is produced from syngas in a methanol reactor. The methanol reactor is a tubular reactor with catalyst loaded into several tubes, surrounded by a bath of boiling water. The boiling water cools the process in an efficient manner and produces steam. The steam is used outside the methanol synthesis unit. The reactor design is such that methanol is synthesised at almost isothermal reaction path and produces conditions for maximum rate of reaction. As a result, there is high conversion per pass and low formation of by-products. Methanol can be co-produced along with generation of electricity from coal-fired power plants also (Fig. 3.11).

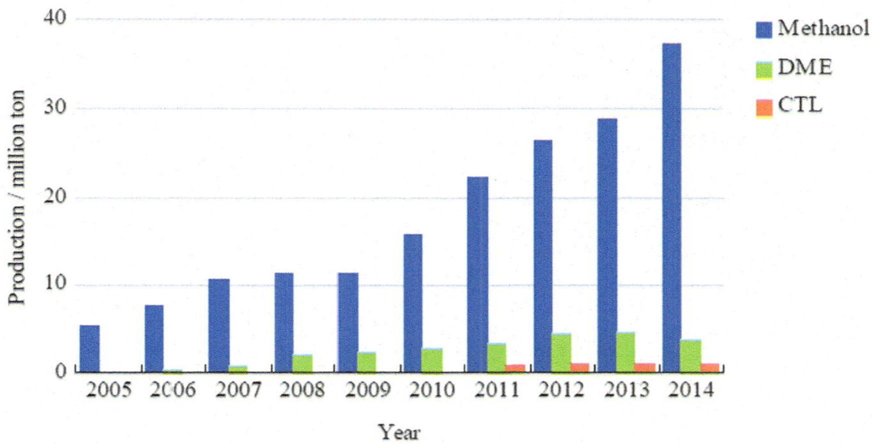

Fig. 3.10 Methanol production in China (China's use of fuel methanol and implications on future energy trends)

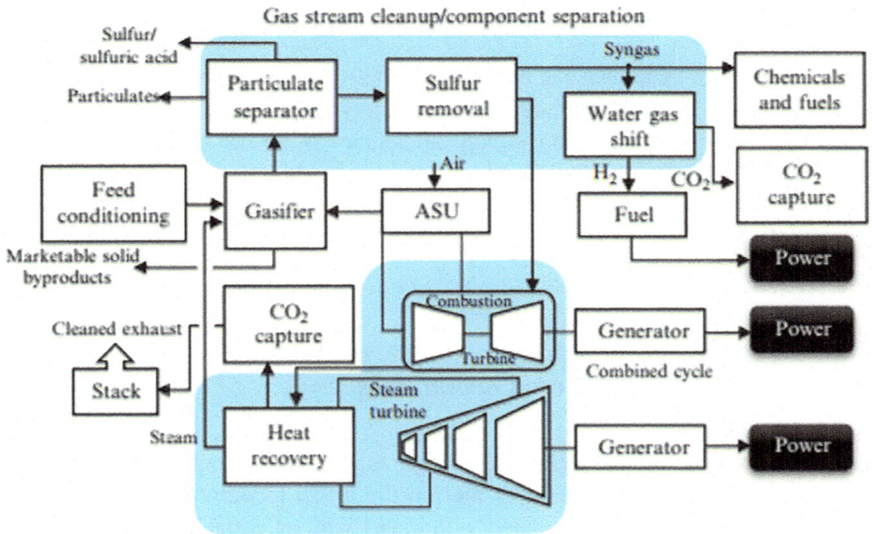

Fig. 3.11 Electricity generation and co-production of methanol in a coal-fired power plant (Ghasemzadeh et al. 2018)

In this process, syngas is produced as an intermediate product. This syngas is then used for synthesis of methanol. The name of syngas originates from its use as intermediates in creating synthetic natural gas (SNG) and for producing ammonia or methanol.

In India, Thermax Global, a Pune-based firm has set up a pilot-scale 'coal gasification to methanol production' plant. BHEL has also set up similar plants at

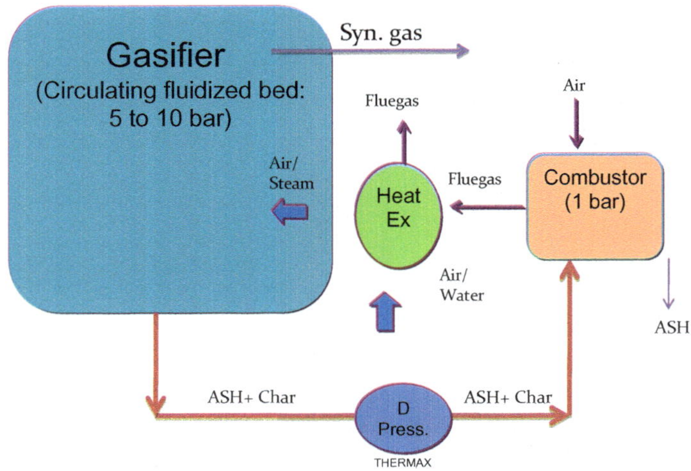

Fig. 3.12 Coal-to-syngas plant schematic, developed by Thermax (2018)

Tiruchirappalli and Hyderabad. The plant design on a very high level is shown in Fig. 3.12.

Thermax plant employs a circulating fluidized bed gasifier technology. The gasifier has been designed to cater to wide variety of fuels, e.g. high ash coal and low-volatile biomass. High conversion efficiencies of >85% in the gasifier have been demonstrated. Energy recovered from unburnt carbon in the ash is realised in a combustor. Thermax claims that overall conversion efficiencies of >95% have been achieved. Gas cleaning system for coal and low tar applications has been developed as part of this plant.

3.4.4 Using Methanol on Locomotives

Methanol is a high-octane fuel and has high auto-ignition temperature. There are various methods to use methanol as fuel on the diesel locomotive engines. Small amounts of methanol can be blended with diesel with an equal amount of suitable surfactant to burn the methanol–diesel blend. In CI engines, methanol can be ignited with the aid of glowplug, since its auto-ignition temperature is quite high. Methanol can be port injected during the intake stroke and stock diesel injectors can initiate diesel ignition directly in the cylinder. Dual injectors or single injector can be used to inject pilot diesel, followed by methanol into the cylinder. Of late, research has been initiated for using ignition improvers and methanol–diesel mixtures directly into CI engines. This research is, however, at a nascent stage. For heavy-duty large-bore engines such as the ones used in locomotives, a single injector to inject pilot diesel and methanol bulk fuel near TDC appears to be the correct technical choice. A financial analysis was carried out to study the economic aspect of transitioning to methanol as a fuel for diesel locomotives of Indian Railways. Financial analysis was done based on the cost estimates and fuel savings projected as given in Fig. 3.13. As the number of locomotive conversions increases, payback period reduces significantly. This is estimated to savings of ~Rs. 2.5 Crore per locomotive per year as a result of switching over to methanol.

Density and calorific value of the two fuels are different; therefore, cost comparison has been done on energy-cost basis. After carrying out the necessary conversions, it is seen that at the current rates, diesel costs ~Rs. 1.6 per MJ energy, whereas methanol is priced at ~Rs. 0.7 per MJ energy. These differentials are expected to increase as methanol production facilities are developed in India to produce methanol from indigenous coal, and the petroleum prices are expected to rise further and become more realistic.

Figure 3.13 summarises the savings to be accrued due to progressively converting the diesel locomotive fleet to be operated using methanol. It is based on the assumption that all diesel locomotives, i.e. both four-stroke and two-stroke, are converted to run on methanol. This conversion exercise can be simultaneously undertaken by DLW and DLMW. At present, India is a net importer of methanol from China because of lower cost of import compared to the indigenous production of methanol. India produces methanol from natural gas imported from Qatar at a very high rate (USD 16 per MMBTU), whereas China produces methanol from its indigenous, cheap high ash coal, leading to a large price differential. Indian methanol costs ~Rs. 23/kg, whereas the imported Chinese methanol costs ~Rs. 16/kg. India

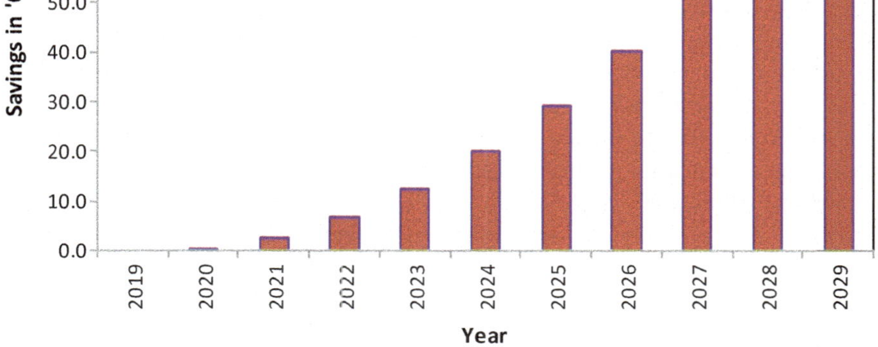

Fig. 3.13 Savings due to adoption of methanol on diesel locomotives of Indian Railways

has already taken up a programme to produce methanol from its indigenous high ash coal. Coal-to-methanol production in India is likely to start in the next three years. At that time, converted methanol locomotives will be fuelled by methane produced indigenously by methanol production units. Thus, for the first 300 methanol locomotives, a price of ~Rs. 23/kg of methanol has been taken, whereas for the balance conversions of 2500 diesel locomotives, fuel price of ~Rs. 16/kg has been considered for the analysis.

It is seen from Fig. 3.13 that there is a saving of more than Rs. 80,000 Crores over a period of 10 years. Considering that methanol is the best medium for storage of renewable energies such as solar, wind, hydro, it makes strong economic sense to manufacturing new locomotives, which are tailored and designed to use methanol.

3.4.5 Methanol-Based Locomotives for Sustainable Railway Traction Beyond Electrification

There have been many inaccurate claims in the media world over that electric vehicles powered by batteries will replace the IC engine powered vehicles. There are also tall claims that whole of electrical grid shifting to renewables by 2035, mainly powered by solar and wind energies. India has planned to set up solar generating capacity of ~175 GW by 2022. Although this itself is a very ambitious target, even if considered as doable, the utilisation load factor (ULF) of solar PV is ~10% due to intermittent nature of source of energy. Therefore, in effect the 175 GW of installed solar capacity translates into only ~20 GW of solar electricity generation by 2022 in the best-case scenario. Also, all solar PV, being planned and used in India for various applications is imported, mainly from China. As a result of very low prices of solar PV panels from China, Indian solar PV panel manufacturing industry is already completely wiped out. According to various studies as cited above, coal would therefore continue to be the main source of electricity generation in India up to 2050 and there are no two ways about it. Considering that India is importing 27.8% of its coal for electricity production (https://coal.nic.in/content/production-supplies) since indigenous coal is high ash content and cannot be used by power plants, India will be dependent on imported fuel for its electricity generation as well, other than importing solar panels from China. This inter alia means that even if whole of Indian Railways is electrified and all diesel locomotives are retired prematurely (leading to a loss of ~Rs. 60,000 Crores to Indian Railways), Indian Railways shall continue to suffer on all the three E fronts, i.e. economy, efficiency and environment.

There is a need to revisit the policy of total electrification of Indian Railways by policy planners. There is an immediate need to look at alternate clean fuels that can use indigenous coal and MSW to produce technically and economically feasible transport fuels. Methanol/DME is one such set of fuels, which can be produced from indigenous coal and MSW, along with cogeneration of electricity, leading to large-scale reduction in emission of pollutants and allowing economies of scale, which should be exploited. This important step shall also reduce our dependence on imported materials/fuels for energy production. The use of methanol either in piston engines, gas turbines or in fuel cells would increase the efficiency of power plants because of its high volatility and lower heat of vaporisation. Efficiency of methanol piston engines increases by ~15 to 20% and can reduce the size of the engine for the same rated power substantially.

3.4.6 Methanol Use in Fuel Cell-Based Traction

Methanol fuel cell hybrids can be developed by Indian Railways, similar to what other leading locomotive manufacturers are developing world over. Efficiency of fuel cells is considerably higher than other types of power plants (Fig. 3.14).

Fuel cell efficiencies are in the range of 30–80%, which can assist in further reduction of fuel consumption and emissions. On-board reformers are used to convert methanol into hydrogen, which is then used in the fuel cells. A schematic of fuel cell locomotive is shown in Fig. 3.15.

In USA, a demonstration fuel cell locomotive prototype has been developed by one of the leading locomotive manufacturers, and the conceptual model of this fuel cell locomotive is shown in Fig. 3.16.

A European company has developed another prototype of hydrogen fuel cell-based train set which is undergoing field trials. It is expected that this train set will be put to commercial use next year (Fig. 3.17).

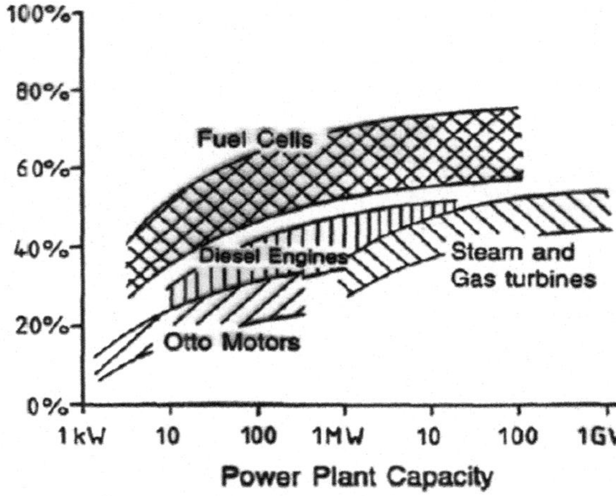

Fig. 3.14 Comparative efficiencies of different power plants (2008)

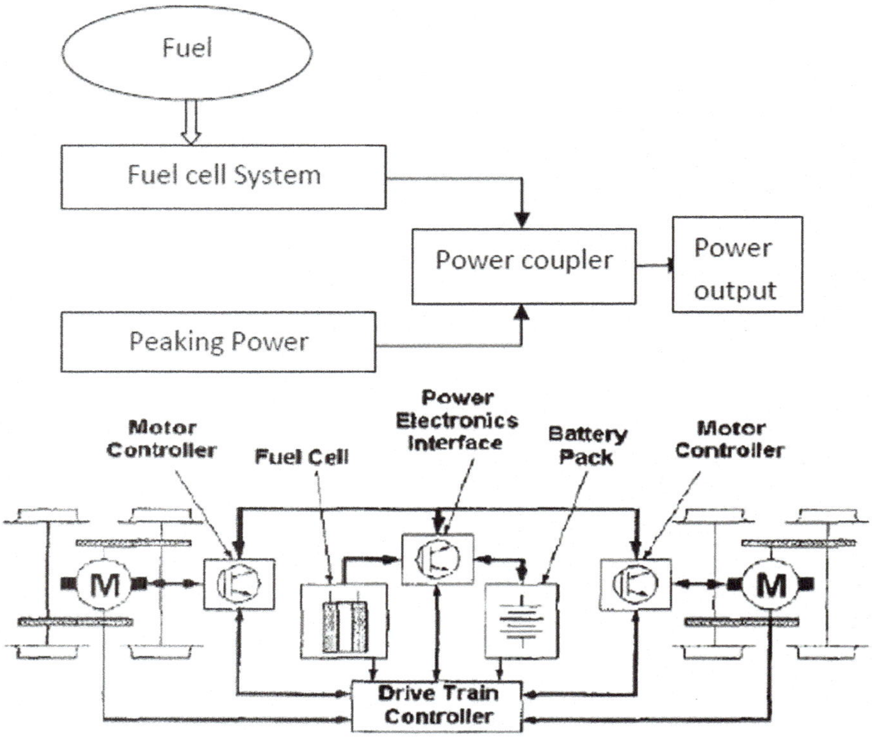

Fig. 3.15 Schematic of fuel cell-powered locomotive (Moghbelli et al. 2003)

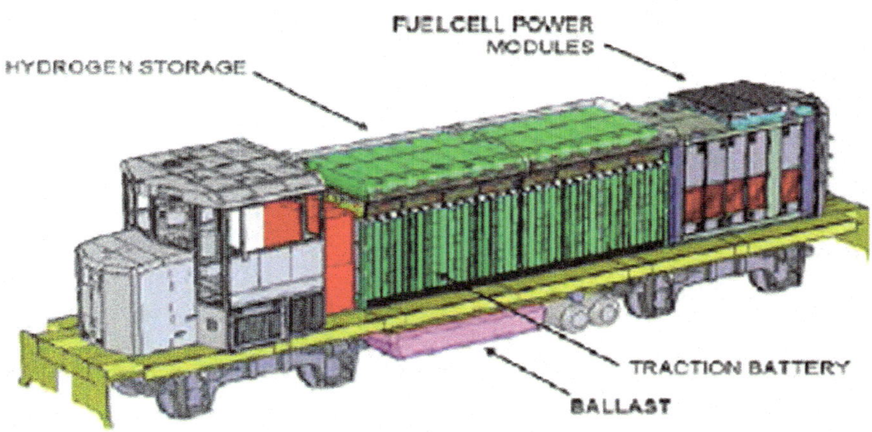

Fig. 3.16 Conceptual model of the fuel cell locomotive prototype (2016)

Fig. 3.17 Hydrogen fuel cell-powered train set prototype developed by ALSTOM (https://www.alstom.com/press-releases-news/2017/3/alstoms-hydrogen-train-coradia-ilint-first-successful-run-at-80-kmh)

Use of methanol coupled with on-board energy storage can help Indian Railways and India to meet the objectives of economy, emissions and energy security simultaneously, and India can sustain itself on indigenous methanol/DME for the next 50–75 years, till the time global transitions into hydrogen economy/solar energy capture and use mature into sustainable and commercial technologies.

3.4.7 Comparison of Different Fuel/Engine Technologies

Table 3.2 compares different combination of fuels/engine technologies for locomotive traction. In this table, four different locomotive technology options, namely (i) diesel engines fuelled by diesel, (ii) electric locomotive fuelled by coal-based electricity, (iii) diesel engine fuelled by methanol and (iv) fuel cell fuelled by methanol, have been compared.

It is observed that electric locomotives with coal-based electricity have the highest GHG emission intensity of 120 g CO_2 equivalent/MJ of energy. This is followed by diesel locomotives using diesel emitting 95 g CO_2 eq./MJ of energy. Diesel locomotives fuelled with methanol show a considerable reduction in the GHG emission intensity of 50 g CO_2 equivalent/MJ of energy and the lowest GHG intensity was exhibited by methanol fuel cell-based locomotives with 10 g CO_2 emission equivalent/MJ of energy. Efficiency-wise figures further corroborate this trend. Methanol fuel cell-powered locomotives are expected to operate in the efficiency range of $\sim 70\%$, whereas the electric locomotives have a very low overall efficiency of $\sim 19\%$. On emissions front, fuel cell-based power plants produce negligible emissions, followed by methanol fuelled diesel locomotive engines.

Table 3.2 Comparison of different fuel/power train technologies for railway traction

	Economics	Environmental impact	Overall energy conversion efficiency (%)	GHG equivalent emissions
Diesel locomotive fleet	Rs. 16,000 Crore/ annum@ 2.8 billion litres of diesel	NO_x-353.6 kton/year PM-15.9 kton/ year HC-25 kton/ year	33	95 g CO_2 eq./MJ of fuel energy
Electric locomotive fleet	Rs. 11,500 Crore/ annum,Rs. 350,000 Crores for electrification infrastructure; Equivalent to 14.8 billion litres of diesel	NO_x-483.2 kton/year PM-80.5 kton/ year SO_x-483.2 kton/year Hg-24 ton/year	19	120 g CO_2 eq./ MJ of fuel energy
Methanol locomotive equivalent fleet with IC engine technology	Rs. 10,000 Crore/ annum, equivalent to 2.2 billion litres of diesel	NO_x-75 kton/ year PM-NIL HC-minimal	38	50 g CO_2 eq./MJ of fuel energy with coal-to-methanol and 20 g with green methanol
Methanol fuel cell hybrid train set with equivalent power as diesel locomotive	Rs. 8000 Crore/ annual expense, equivalent to 2 billion litres of diesel	NO_x-NIL PM-NIL HC-NIL	70	10 g CO_2 eq./MJ of fuel energy

Electric locomotives produce the highest specific GHG emissions, mainly due to primarily coal-based electricity generation in India. From the above study, we can safely conclude that methanol-based CI engines and methanol-based fuel cell-powered locomotives are the best choices for Indian Railways for achieving the objectives of economy, efficiency and environment. Table 3.3 compares different parameters for different fuels. Biomass, alcohols, hydrogen and battery/electricity have been examined on different four dimensions, as energy carriers. These dimensions are (a) energy density, (b) storage costs, (c) transport costs and (d) environmental impact.

Battery/electricity performs poorly on all four accounts, whereas alcohols stand out for exceptionally good performance on these four counts. This study underscores a strong case for methanol adoption on Indian Railways as a traction fuel.

Table 3.3 Comparison of different energy carriers (Shih 2017)

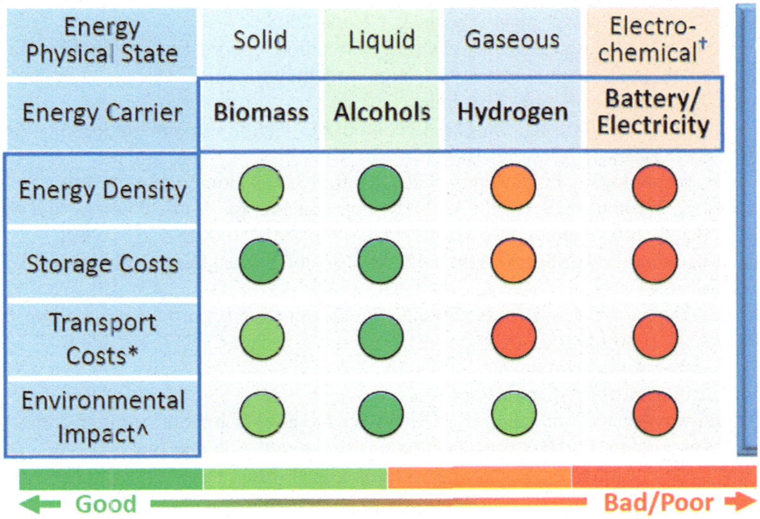

3.5 Conclusions

There is a growing market for diesel locomotives and diesel vehicles worldwide, with a growth rate of >3%. Chinese Northern Railway (CNR) has established itself as a leading manufacturer and exporter of diesel/self-powered locomotives and rolling stock worldwide. Indian Railways has also created an indigenous industry and ecosystem around diesel/self-powered locomotives and rolling stock units in the last half century. For reasons of self-reliance and energy security, Indian Railways should adopt methanol as a fuel to power its diesel locomotives. Indian Railways must also invest in multiple traction technologies and fuels for achieving the goals of economy, efficiency and environment. It is already too late for Indian Railways and RDSO to start working on hybrid powertrains for locomotives and fuel cell-based traction technology for railway propulsion. There is a strong case for Indian Railways to develop indigenous technologies in these areas, considering their usefulness and potential for export in the global market.

References

Abbott D (2010) Keeping the energy debate clean: how do we supply the world's energy needs? IEEE, New York

Chen JS (1995) The production of methanol and hydrogen fuels from municipal solid waste. Thesis for MS from Princeton University

China's use of fuel methanol and implications on future energy trends, Natural Gas Markets Team, U.S. Energy Information Administration

Comparison of different power plant efficiency T.Huria 2008

Das N, Roy J (2018) Making indian power sector low carbon: opportunities and policy challenges. In: Gautam A et al (eds) Sustainable energy and transportation, energy, environment and sustainability. Springer Nature, Singapore

Draft Technology Assessment, Freight Locomotives, California Environmental Protection Agency, Air Resources Board, 2016

Future Power Technologies, Final Report, CRC for Rail Innovation, 2013

Ghasemzadeh K, Tilebon SMS, Basile A (2018) Coproduction of electrical energy and methanol in IGCC Plants. https://doi.org/10.1016/B978-0-444-63903-5.00015-7 (Elsevier)

Government of India, Power Sector, Jan-2017, Ministry of Power, Central Electricity Authority, New Delhi

https://www.alstom.com/press-releases-news/2017/3/alstoms-hydrogen-train-coradia-ilint-first-successful-run-at-80-kmh

https://www.automotive-iq.com/tag/adas

https://coal.nic.in/content/production-supplies

http://www.railwaygazette.com/news/single-view/view/ge-unveils-hybrid-locomotive.html

https://www.sae.org/news/2016/08/sustainable-transformationthe-daimler-wy

https://www.technavio.com/report/global-powertrain-global-diesel-locomotive-engines-market-2016-2020

India's intended nationally determined contribution: working towards climate justice, 2015. http://www4.unfccc.int/ndcregistry/PublishedDocuments/India%20First/INDIA%20INDC%20TO%20UNFCCC.pdf

Moghbelli H, Gao Y, Langari R, Ehsani M (2003) Investigation of hybrid fuel cell technology applications on the future passenger railroad transportation. In: Proceedings of the 2003 IEEE/ASME joint rail conference, Chicago, 22–24 April, pp 39–53

Perspectives for the energy transition Investment Needs for a Low-Carbon Energy System, IEA (2017)

Shih CF (2017) Opportunities and pathways to a green future for all. Opportunities and challenges for methanol as a global liquid energy carrier. Stanford University, July 31st–August 1st 2017

Thermax Presentation in the 6th meeting of Methanol Engine Task Force on 12th July, 2018 at Lucknow

Todd Davidson F, Nagasawa K, Webber ME (2017) Electrofuels. ASME Mechanical Engineering Magazine, Sep 2017

www.irena.org

www.eia.gov

Chapter 4
Enabling Rural Economy in India to Partially Substitute Petroleum Products by Methanol—Technology Solutions and Policy Support

M. S. Srinivasan

Abstract An estimate of the energy for cooking and transportation needs of rural India is made, and it is established how there would be a serious pressure on imports if such needs are attempted to be served using only conventional petro-products. Methanol derived from biomass could be one way to partially substitute the petro-products in rural areas. While there is a need to step up efforts for indigenous technology, especially for the gasification of such biomass, it is argued that it is more important to formulate a consistent policy to support both the technology development and its market adoption. A road map to set up modular bagasse-based gasification cum methanol plants in a distributed manner across the country is described. In order to make these plants commercially viable, policy support in terms of initial capital subsidy and an administrative pricing mechanism is suggested.

Keywords Biomass gasification · Viability of methanol production

4.1 Introduction

As one looks at India as she moves closer to the 75th year after regaining independence, there are several matters for considerable satisfaction—rising education level, access to healthcare, self-sufficiency in food production and political stability with vibrancy. Most of these improvements have been on account of introduction and implementation of policy measures by the Government of India in response to prevailing social conditions and a determination to pursue desired developmental objectives, going beyond conventional wisdom.

M. S. Srinivasan (✉)
Indian Institute of Technology, Madras, Chennai 600036, India
e-mail: iro@wmail.iitm.ac.in

A. K. Agarwal et al. (eds.), *Methanol and the Alternate Fuel Economy*, Energy, Environment, and Sustainability,
https://doi.org/10.1007/978-981-13-3287-6_4

One aspect of social policy that requires critical examination at this juncture is the framework to provide for clean and secure energy, especially for the rural sector, and doing it in an economically and environmentally sustainable manner. This paper sets out one such approach by outlining certain policy measures to enable valorization of agri-residues using well-known technological principles. This would supply a portion of the growing energy needs for fuel and transportation in rural India.

4.2 Analysis of Energy Needs—Rural and Urban India

4.2.1 Qualitative

At present, there is a sharp difference between the urban and rural population with respect to the source of energy used for cooking and transportation purposes. Urban use is mostly based on LPG for cooking and diesel/petrol for transportation. Rural use, on the other hand, is largely kerosene/biomass for cooking and diesel/petrol for transport. The National Sample Survey Office report 68th round brings this out vividly.[1] Two-thirds of rural households use firewood and chips for cooking.

The decreasing availability of such biomass and the risk to the health of housewives and children exposed to the smoke are matters of concern prompting the government to introduce schemes like Pradhan Mantri Ujwala Yojana (PMUY) to make LPG available at subsidized rates to rural poor. Laudable as this is, however, there is a problem of large and expensive imports due to this shift. Similar is the situation with respect to transport fuels. A survey by Nielson done on behalf of the Petroleum Planning and Analysis Cell in 2013 showed that at all India level, almost 60% of the retail sale of petrol is for consumption needs of two wheelers.[2] In terms of rural–urban consumption, again the rural areas are growing at a faster clip in terms of two-wheeler sales, thus adding to the increasing need for petrol in rural areas. One also has to keep in mind that diesel consumption is very strong in rural areas to power agricultural tractors, pump sets, etc., and this too is likely to increase in future. An estimate is made in the next section on the likely additional consumption needs of petro-products in rural India.

[1]Energy sources of Indian households for cooking and lighting, NSS 68th round, July 2011–June 2012.

[2]All India study on sectoral demand of diesel and petrol, Petroleum Planning and Analysis Cell, 2013.

4.2.2 *Quantitative Estimates*

4.2.2.1 Cooking Needs of Rural Areas

If all 17.2 Crore rural households (as per NSSO report) were to switch to LPG for their cooking needs and assuming a monthly consumption of 10 kg per household, the *additional LPG requirement will be 17.5 million tons/year.* This has to be seen against the all India LPG consumption during 2017–18 of 23.5 million tons of which nearly 50% is imported. As India has limited domestic production of crude oil, it is almost certain that such quantities will have to come exclusively from imports.

4.2.2.2 Personal Transportation and Use of Agri-Implements Needs of Rural Area

In the absence of field data on the percentage of total petrol that is consumed in rural areas, an assumption is made that one-third is thus consumed. Thus, out of the 10 million tons of petrol, *we could approximate 3 million tons per year in rural areas.* To this, one needs to add the diesel consumption of agricultural implements. The Petroleum Planning and Analysis Cell report of 2013 indicates that about 14% of diesel is used for non-transport needs like agricultural implements/generators, etc., based on which we could estimate that the *annual diesel consumption for such use is approximately 2.7 million tons. At a growth rate of 5%, the additional annual requirement of diesel can thus be estimated as 0.14 million tons.*

4.3 Affordability

We will begin with an analysis of the cooking fuel. The NSSO survey report 558 indicates that during 2011–12, the average monthly expense of a rural household on fuel is about Rs. 60.[3] The current subsidized price of a 14.2 kg cylinder of LPG is Rs. 495. If we take that the monthly consumption of LPG for a family of six people is 10 kg, the cost of switching to LPG for the household is Rs. 290 per month. Thus, there is a sharp increase in costs even at the current subsidized pricing of LPG. The subsidy on a LPG cylinder varies of course with the weighted average price of crude, and the current (Aug 2018) subsidy that government extends is Rs. 250 per cylinder or Rs. 18/kg adding to government funding the rural household of about Rs. 180 per month.

Thus, it can be seen that if we would like to achieve the social goal of clean fuel to the rural poor, there are additional costs that both the consumer and the

[3]Household consumption of various goods and services, 2011–12, NSS 68th round.

government have to absorb. For the government, for the additional 17.5 million tons of LPG, this is a hefty Rs. 31,500 Cr/year. If we consider the incremental petrol and diesel consumption in rural areas, it is recognized that the subsidy on these fuels is almost phased out considering the use is by consumers who are not in lower income category. Nevertheless, the import dependence would continue. A final word on this analysis is important. As the international crude prices go up, the additional costs also would go up. More critically such a flow of money would be out of the country with no contribution to local economy.

4.4 An Alternative Approach—Use of Biomass to Deliver Clean Energy in the Form of Methanol to Rural Households

Use of methanol in a specially designed stove for cooking is already well demonstrated (https://stoves.bioenergylists.org). Also, use of methanol blends in both petrol and diesel as a co-fuel in IC engines is also well established. Thus, methanol could be considered to substitute LPG as a cooking fuel and partial substitute for petrol and diesel in IC engines.

At a global level, methanol is produced primarily from natural gas, and this again is not available within in India sufficient quantities. However, biomass, more particularly bagasse from sugar cane production, is available in the country. Therefore, at the conceptual level, an approach based on converting available biomass to syngas and onto methanol is worth exploring in detail. This paper will take sugar cane bagasse as a typical biomass for further illustration.

4.4.1 Making Syngas Using Biomass and with Low CO_2 Emission and Converting into Methanol

There are well-documented works in the scientific literature on gasification of biomass with air/steam mixture. The carbon conversion efficiency and cold gas efficiency have been studied well (Riehl et al. 2012). Most of these studies have been on a small scale where the raw material feeding and gas treatment systems were rather elementary as the focus was mainly on the gasifier. Such efforts have considered burning of the syngas thus produced in a gas turbine, and accordingly it is the economics of gas turbine-based power generation that has dictated if a larger-sized demonstration plant could be built. Obviously under the prevailing economic models of power generation, such efforts did not lead to viability. Also the large-scale collection of biomass was found to be not feasible.

However, not much work seems to have been done on *oxy/steam reforming* of bagasse. Oxyfeed is essential to avoid dilution with nitrogen from air that would be unsuitable for methanol production. Further, the effective cleaning and conditioning of syngas is critical for methanol, and no reliable data is available on this aspect. Therefore, there is a need for setting up a demonstration plant for gasification of biomass like bagasse in an oxy/steam environment to make syngas fit for fuels like methanol and other alcohols. The parameters relating to optimizing the cold gas efficiency and gas cleaning/conditioning are to be established in such a demonstration plant. Both to minimize efforts at further scale-up and in view of the rather distributed availability of the rather bulky raw material, it is necessary that we aim at a demo plant of a size that can accommodate the total raw material available at a district level. It is estimated that such a gasification plant would be of approximately 60 tpd methanol equivalent.

From practical considerations of high reliability in operations, it is envisaged that the gasifiers be built in a configuration of 3*50% capacity units. After successful establishment and parameterization, a suitable single-stream methanol plant may be attached to the gasifier block. This approach is suggested since the methanol portion of the technology is mature and well developed and may not require any demonstration scale efforts to be redone. Setting up a methanol plant of smaller size is recommended after the generation of proven and reliable gasifier design. The conventional wisdom is that only very large methanol plants of 2500 tpd and above are likely to be viable. However, this logic has evolved using natural gas as feedstock and has to be critically re-examined for a feedstock like bagasse. This aspect will be elaborated in a later section.

Another question that is frequently asked if a methanol plant based on a non-natural gas feedstock does not add to the carbon footprint. This is asked in the context of import of either methanol or natural gas as opposed to manufacturing from biomass. It is acknowledged that the CO_2 emission from a conventional biomass-based methanol plant is higher. Therefore, along with the development of gasification technology, it is also suggested that an effort is made to modify the methanol flow sheet to supply a part of the endothermic energy needed by the gasifier through a form of solar energy. This could be taking the form of a high-temperature solar furnace or addition of solar-generated hydrogen to adjust the CO_2 balance (https://cse.umn.edu).

4.4.2 *What Is the Likely Methanol Plant Size Based on Biomass?*

The starting point is the recognition that biomass-based methanol plant size would be dictated both by the availability of biomass and the need to minimize transport of bagasse/methanol over long distances. Thus, a concept of one methanol plant for every district is a reasonable starting assumption. Beginning with consumption: as stated earlier, methanol could be used as a substitute for LPG for cooking. One could estimate the requirement of methanol as follows:

- The average number of rural households in each district is approx 300,000 nos. If about 20% of such households would shift from firewood to methanol stoves, we have 60,000 households in each district. At estimated daily energy consumption for cooking of 15.8 MJ/day/household, the annual additional methanol requirement is estimated as (60,000*350*15.8/15.78/1000) KL per year. This works out to 21,000 KL or 16,600 tons per year, for a district.
- To this we could add the methanol needed for co-firing of diesel-fired tractors and two wheelers in rural areas running on petrol. The current all India petrol consumption accounted by three wheelers approximately 3 million tons, and at 10% blending by methanol in rural areas, one can estimate this demand at 3 lakh tons/year. For an average district, this equates to 500–600 tons/year.
- It could thus be seen that about 20,000 tons/year is a good starting estimate of methanol consumption in a district.
- A 20,000 tons/year methanol plant would equate to a 60 tpd methanol plant.

4.4.3 Is Adequate Raw Material Available?

Mass balance calculations show that approximately 4 tons of "as is" bagasse with approximately 50% moisture content is needed for 1 ton of methanol output. Hence, for a 20,000 t/year methanol plant, we require approx. 80,000 t of bagasse. As per All India Sugar Mills Association data, there are 493 sugar mills processing an expected 400 million tons of sugar cane in 2017–18. This gives an estimate of availability of more than a lakh ton of bagasse per sugar mill. Every district of our country has a few sugar mills. At present, the bagasse is utilized by the sugar mills for cogeneration of power, and the availability for a methanol plant will depend on the possibility of getting a higher realization than at present. This point will be elaborated under a later section.

It therefore seems that subject to a suitable price discovery, availability of bagasse to support a district-level methanol plant is unlikely to be a problem. Further to enhance availability of more biomass, as part of gasification design, the following two materials as cuttings from sugar cane stalking and paddy straw should also be considered viz raw materials for gasifiers. There are already reports of such material being successfully demonstrated in air gasifiers.

4.4.4 Is a Small Methanol Plant Economically Viable?

A plant's viability depends on the selling price and the cost structure. So it would be for a methanol plant based on bagasse. Let us examine the selling price carefully. In view of the increasing integration of India into the global economy, such economic analysis has to be done in the global context.

4.4.4.1 International Methanol Prices

This is determined largely by the very large plants of 3000–5000 tpd capacity based on natural gas and set up in the Arab Gulf countries. Such plants enjoy several cost advantages viz economies of scale, ease of processing of natural gas that lead to lower capital and operational cost and most important of all, the very low local price at which natural gas is made available to such plants. In terms of quantitative parameters, these advantages can be measured in terms of:

- estimated capital cost per ton of methanol—about $500 per ton per year,
- energy consumption for feed and fuel per ton methanol output—7.9 Gcal/ton,
- feed gas price—$1–2 per Gcal.

With such advantages, the cost of production of such plants is lower. An estimate shows the cost of production of ex-plant gate from such a plant is likely to be between $75 and $110 per ton. The selling price as it applies to India has fluctuated between $200 and $650 per ton on CIF basis.

4.4.4.2 A Natural Gas-Based Methanol Plant in India

There are a few methanol plants in India both in the public and in the private sector. The feedstock has been a mixture of natural gas, naphtha and fuel oil. These plants are all technically well proven but have been unable to run to capacity mainly due to price competition from imports. The disadvantages of Indian plants have been:

- higher feedstock price—more than twenty times as compared to the plants located in the Middle East,
- slightly higher energy consumption, especially for liquid fuel-based plants,
- higher capital costs in view of much smaller plant sizes—in the sub 500 tpd range.

4.4.4.3 Situation with Respect to a Biomass-Based Smaller Methanol Plant

As earlier proposed under Sect. 4.4.1, small methanol plants based on bagasse would have to face all three handicaps as compared to the mega-plants in the Middle East. In particular, the feed and fuel requirement is likely to be much higher. A well-designed demonstration plant would establish the actual numbers, but it appears that there would be handicap of at least 40% on this account, thus giving an estimate of 11 Gcal/ton methanol produced.

Also the gasifier and gas cleaning system will be more complex as opposed to a conventional natural gas fed steam reformer; we can expect that the capital investment will thus be higher. The economies of scale also would operate unfavourably.

Against this, there is the possibility of a larger choice of equipment vendors since the range of equipment involved could be made by a potentially larger group of qualified vendors. The higher volume of business for capital goods industry (one plant per district means almost 400 plants!) is also a favourable factor. Nevertheless, the per unit capital investment is certain to be higher. A demonstration plant will provide an indication of the eventual costs, but taking a multiple of 2 seems to be a good starting point.

That leaves the issue of pricing of bagasse. At present, the price discovery of this material is based on generation of electrical power in a cogen unit attached to the sugar mill. The purchase price of electricity determines the opportunity price of bagasse. A detailed study of this aspect across the country is needed, but from the information available from a large sugar mill in Tamil Nadu, a price of Rs. 1600 per ton of as is bagasse with 50% moisture has been imputed based on the purchase price of electricity sold to the Grid. At a calorific value of 2.3 Gcal/ton of such material and using this price discovery, we can arrive at Rs. 696 per Gcal of bagasse —equivalent to approximately $10/Gcal. In comparison, as pointed out earlier the natural gas is made available to the Middle East based methanol plants at $1–2 per Gcal! As one is looking for value addition for bagasse, there seems to be little possibility of a lower price in future.

4.4.4.4 Summary

Clearly in addition to technology efforts, policy support is needed to make the concept of biomass-based methanol plants viable in India based on non-petroleum feedstocks. It is emphasized that the above conclusions are valid even if we consider coal as a feedstock. For example, for a coal of 5000 kcal/kg calorific value and using a pit-head price of Rs 3000 per ton, one can arrive at energy cost of Rs. 600 per Gcal or equivalent to $7 per Gcal. In conclusion, it is necessary to appreciate that a small capacity methanol plant in India can be made viable only if suitable policy support is extended. *No amount of technology development, however welcome, is likely to compensate for this requirement.*

4.5 Policy Support—Few Thoughts

It was pointed out in the Introduction section that India has had an outstanding record of conceiving and implementing policy measures that have significantly contributed to development with equity. The programme of methanol production using Indian feedstocks to partially substitute imported petro-products is another opportunity for coming out with a similar policy. Unlike the past, however, the present global situation is more constrained in terms of WTO requirements under which a country affected adversely by what it perceives as "free access" of its products in another country could raise a dispute in the WTO. As mentioned earlier, there is an inherent price differential between cost of energy in a few selected countries located in the Middle East and in India. In the opinion of the author, the pricing of natural gas in exporting countries itself is a subject and could provide grounds for the introduction of a well-thought-out policy. A few suggestions are offered in this context:

- A one-off capital subsidy based on licence-free use of indigenous technology of gasifiers to companies willing to set up methanol plants attached to sugar mills. This could be on the basis of open auction.
- A price protection for about 10 years by safeguard import duties on imported methanol. The oil marketing companies may be supplied methanol at such import parity prices.
- To compensate for the additional costs at consumer level, special subsidy schemes that are designed to taper off in relation to international crude prices may be designed for methanol supplied to cooking needs.
- Such costs have to be seen against the potential subsidy cost indicated under Sect. 4.3.

4.6 Conclusions

The title of this paper specifically mentions rural economy in the context of partially substituting petro-products using methanol generated from indigenous sources. A word of clarification may be in order to explain why this specific reference. If we start with a goal of providing clean, safe sources of energy to rural population that are sustainable and also offer some security against International uncertainties, it is important to recognize that any solution shall have to be replicated across the country. The model of a few centrally located producing units distributing the finished product across the country has its limitations. For using agro-raw materials like bagasse, this issue is very critical.

A centralized solution is probably impossible and for blending with petrol/diesel too the logistics are complex. Thus, it may be helpful to do this substitution at a local level, and this justifies the reference to rural economy. In conclusion, in this paper, the technical status of use of agri-based raw materials such as bagasse to make methanol is briefly mentioned. It is highlighted that in addition to encouraging technology development, it is equally important to understand the issues of making such an initiative viable by policy support. A few suggestions are offered for wider debate and for arriving at suitable policy measures.

Acknowledgements The author thanks the Dean ICSR, Indian Institute of Technology, Madras (IIT-M), for generous support and encouragement for this work. Thanks are also due to Prof. A. Ramesh of IC Engines laboratory and Prof. S.R. Chakravarthy of National Centre for Combustion Research and Development (both at IIT-M) for stimulating discussions.

References

A ray of sunshine for bio energy, College of Science and Engineering, University of Minnestoa, https://cse.umn.edu

Commercialisation of a new stove and fuel system for household energy in Ethiopia using ethanol from sugar cane residues and methanol from natural gas, Project Gaia, https://stoves.bioenergylists.org

Riehl RR, Shahateet CA, de Souza LS, Karam D Jr, Ambient N (2012) Biomass gasification unit using sugarcane bagasse for power generation. In: Recent Advances in Energy, Environment and Economic Development, pp 222–227. ISBN: 978-1-61804-139-5

Part III
Alternative Fuels

Chapter 5
Study of Performance and Emissions of Engines Fueled by Biofuels and Its Blends

Gaurav Dwivedi, Suyesh Pillai and Anoop Kumar Shukla

Abstract Since ages, fossil fuels have been used for the purpose of running engines and machines. Due to increasing concerns regarding the use of fossil fuels like greenhouse gas emissions, renewability, sustainability, increasing population, and high fuel costs, the focus has now shifted on to the use of biodiesel as an alternative fuel. Biodiesel can be easily derived from edible as well as non-edible and algal material. It is renewable, sustainable, and cost-effective source of energy. Use of biodiesel can effectively increase the performance of engines and reduce emissions due to which major energy-consuming nations are producing biodiesel and using it in blends with other fuels. This paper focuses on the study of advantages and disadvantages of using biodiesel as a fuel, sources as well as method of biodiesel production, global energy scenario, fuel properties of biodiesel and its comparison with other fuels, and the performance and emission characteristics of engines whenever they are fueled with various fuels and additives.

Keywords Emission · Methyl tert-butyl ether · Additives · Brake thermal efficiency

5.1 Introduction

Since about 100 years, petroleum fuels derived from fossil deposits have been used to drive and run automobile vehicles and perform industrial production. In another 20 years, the requirement for energy will be almost 50% more than the current usage due to increasing population and industrialization (Kumar Vijay et al. 2017). On the other hand, globally, the sources of petroleum-based fuels are decreasing day by day. Also, use of these fossil fuels leads to large amount of greenhouse gas

---

G. Dwivedi (✉)
School of Mechanical Engineering, VIT University, Vellore, India
e-mail: gdiitr2005@gmail.com

S. Pillai · A. K. Shukla
Amity University Uttar Pradesh, Noida, India

A. K. Agarwal et al. (eds.), *Methanol and the Alternate Fuel Economy*,
Energy, Environment, and Sustainability,
https://doi.org/10.1007/978-981-13-3287-6_5

emissions which is expected to increase by over 80% if the usage of fossil fuels increases in next 20 years. Also increasing price of fossil petroleum by salient drivers has come up as a matter of concern. Hence, the worldwide concern regarding increasing pollution and decreasing fossil fuel reserves causing environmental pollution and also affecting sustainability has led to many research workers working hard to find an alternative fuel obtained from non-petroleum source which will be cost-effective, efficient, and sustainable and produce lesser emissions (Saravanan et al. 2018). Figure 5.1 shows the primary energy supply of the world.

Large reduction in emission of CO makes biodiesel an attractive option for use as alternative fuel. Hydrocarbons (HC), carbon monoxide (CO), and smoke emissions are largely reduced wherever biodiesel is used as fuel in CI engines. Also, ready availability, renewability, and good antiknock qualities make biodiesel a good alternative to fossil diesel. One major disadvantage of using biodiesel is the formation of oxides of nitrogen (NO_x) due to the presence of oxygen (O_2) in biodiesel. Complete combustion of this O_2 does not take place in biodiesel, and hence, NO_x is formed (Abdullah et al. 2017). Different types of biodiesel are available based on the various raw materials that can be used for biodiesel production for, e.g., bioethanol, Pongamia, castor (Ahmad et al. 2011).

Large number of research works has been done to date on the use of biodiesel and its blends as use in various types of engines. Studies concluded that with increase in percentage of biodiesel, specific fuel consumption (SFC) increases as a result of low heating value of biodiesel. The results found in case of emissions were that there is significant decrease in smoke, HC, and CO emissions for all

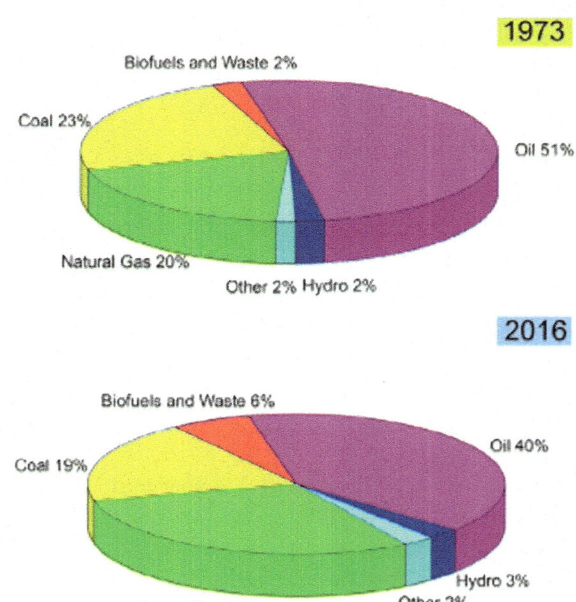

Fig. 5.1 World total primary energy supply from 1973 to 2016 (Saravanan et al. 2018)

proportions of Pongamia biodiesel blends. On comparison of Pongamia biodiesel NO_x emissions with NO_x emissions of diesel, it was found that NO_x emissions are more in case of Pongamia biodiesel than diesel. Devan et al. used 15% of neat poon biodiesel and found that there was significant increase in smoke emissions and improvement in brake thermal efficiency (BTE). Further comparative studies showed that methyl ester gives better performance and produces lower emissions than ethyl ester and diesel fuels (Ashnani et al. 2014). In the present situation, many additives are used in biodiesel to improve quality of fuel, combustion and reduce emissions. The selection of additives that are to be used in order to improve performance is done taking into consideration the various drawbacks of biodiesel, for example, viscosity, density. Metal-based additives are used in biodiesel to reduce harmful emissions as well as to remove unburnt hydrocarbons. The presence of copper(II) chloride ($CuCl_2$), iron(III) chloride ($FeCl_3$), copper(II) oxide (CuO-nanostructured), and cobalt(II) chloride ($CoCl_2$) is what makes these additives useful for reducing harmful emissions. The most used additives along with diesel as well as biodiesel are metal-based additives as they effectively increase the octane rating and also help in better combustion process. The fact that oxygenated additives can be mixed in any ratio with biodiesel without separation of its phases makes it an attractive additive for biodiesel. Even though these additives result in good combustion, still NO_x emissions are produced. Ethanol, butanol, methanol, and propanol are the alcohols that are used as oxygenated additives, whereas di-iso-propyl ether, dimethyl ether, ethyl tertiary butyl ether, methyl tert-butyl ether, and diethyl ether are the most common ethers that are used as oxygenated additives. Esters are also used as oxygenated additives. They include acetoacetic esters, dicarboxylic acid esters, and dimethyl carbonate esters. Hence with the use of oxygenated additives, the fuel will burn more completely, ignite efficiently, and reduce pollution. In engines run on diesel fuel, delay period is usually deiced by the chemical and physical properties of the diesel fuel. Cetane number improvers are used along with diesel fuel to make ignition delay period shorter which will result in better combustion and reduction of unwanted pollutants formation and reduced engine noise. Many research works have been done in order to increase cetane number of diesel fuels as a result of which many additives have been developed. Aldehydes, nitrites, tert-zoles, peroxides nitrates, and alkyl nitrates like amyl nitrate, mixed cetyl nitrates, and hexyl nitrates are most commonly used cetane number improver additives with the alkyl nitrates being the most commercially used additive. The very first effective cetane number improver additive that was used was di-tertiary butyl peroxide (DTBP). This additive was also proved to efficiently reduce even NO_x emissions. Hence, cetane number additives are important to be used along with diesel and biodiesel so as to reduce engine noise, reduce harmful emissions, and increase fuel consumption that is improved combustion. Another class of additives used along with biodiesel fuels is antioxidant additives which are used in order to reduce corrosion darkening color and slush deposits. Butylated hydroxyl toluene (BHT), tert-butyl hydroxyl quinone (TBHQ), butylated hydroxyl anisole (BHA), pyrogallol (PL), propyl gallate (PG) are the most commonly used antioxidant additives among which TBHQ, PG, and PL are

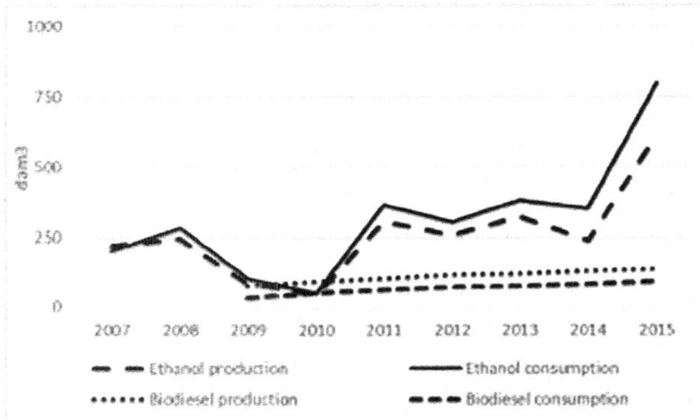

Fig. 5.2 India biofuel production and consumption (Beckman et al. 2018)

the most successful additives. The main purpose of using these types of additives is to get improved stabilizing potential (Ayodele and Dawodu 2014; Babadi et al. 2016; Dawodu et al. 2014; Conceição et al. 2007; Dantas et al. 2007). One of the main reasons for developing biofuels was to decrease greenhouse gas emissions so as to reduce global temperature and avoid the adverse effects of climate change on the ecosystem. Through various experiments and investigations, it is proven that bioethanol and biodiesel are the best alternative fuels for petroleum- and gasoline-derived fuels. Due to the fact that biodiesel can be efficiently produced from animal fats and natural vegetable oils, biodiesel is also given name "mono-alkyl" (Demirbas 2008). Biodiesel creates almost nil emissions of sulfates and aromatic compounds. Even though renewable sources of energy like wind, solar, hydropower have potential to be used as good alternative sources of energy, yet biodiesel received worldwide attention because of its compatibility to be used in engines like CI engine, SI engine which can only be run on diesel fuel. Figure 5.2 shows the production and consumption of biofuel in India.

5.2 India's Energy Scenario

The development of substitute biofuels plays a major role in development of the nation especially in a situation where all the energy needs of the nation are met by the depleting fossil energy (Falade et al. 2008). The use of these fossil energy sources increased 20-fold in the past hundred years due to which there has been adverse climate changes and large amount of foreign import of fossil fuels. If substitute biofuels are used instead of fossil fuels, it will reduce the nation's dependency on large level of foreign import of fossil fuels. In addition to avoiding adverse impacts of climate change on the environment, the production and usage of biofuels can create several job opportunities which would help in the economic developments of the nation (Freire et al. 2009; Masjuki and Abul 2013).

In addition to avoiding adverse impacts of climate change on the environment, the production and usage of biofuels can create several job opportunities which would help in the economic developments of the nation (Freire et al. 2009; Masjuki and Abul 2013). Figure 5.3 shows the fuel projection in India and China from 2008 to 2020 (Beckman et al. 2018).

Because of the increasing demand coupled with depleting fossil fuel resources, the nation's interest is toward using substitute biofuels instead of fossil fuels. At current, biofuels are imported by India from different foreign countries. Efforts are to make use of non-edible sources for the production of biodiesel so that it can reduce the nation's dependency on foreign import. Owing to this mission of producing biodiesel within the country itself, Jatropha curcas, a non-edible oil feedstock, has been grown in the country over four thousand hectares of agricultural land so as to achieve a target of 3.75 tons of biodiesel production per year. This is a major leap toward India's national biodiesel mission which aims at replacing 20% of the fossil fuel diesel usage by biodiesel that is to make use of B20 biodiesel blend (Saravanan et al. 2018; Knothe 2010; No 2014). Karanja has also been proved to be a good source of biodiesel production in India. Jatropha and Karanja were used to produce biodiesel in 200 districts across 18 states in India. Even though India has been constantly and efficiently increasing its biodiesel production, still the consumption of fossil diesel has been increasing due to large population and due to new and developing industrial sectors (Koyama et al. 2007).

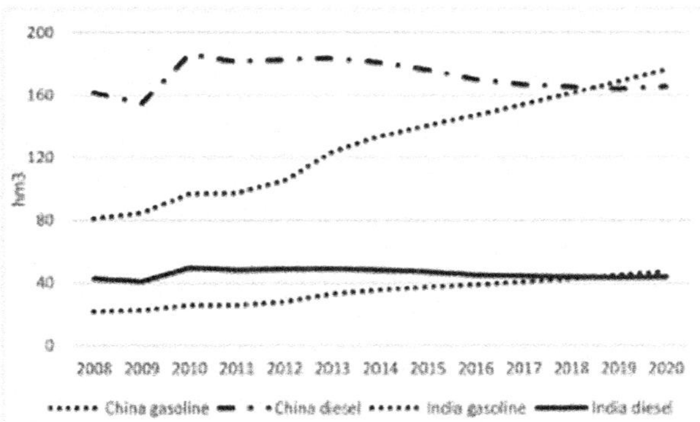

Fig. 5.3 Fuel projection from 2008 to 2020 (Beckman et al. 2018)

5.3 Global Energy Scenario

Global energy scenario refers to the study of performance and consumption of fuel energy by the entire human civilization in this world. With the development of new economies, the global consumption of energy increased by about 840,000 b/d with light distillates being the most rising refined product category. The world energy consumption is expected to increase by about 1.75 × 109 tons of oil by the year 2034, where the current status of utilization at global level is 1.2 × 109 tons of oil. Since India and China alone are two of the most major consumers of energy, these two countries itself are expected to increase the energy consumption of Southeast Asia by up to 75% by the year 2030. Globally, the sources of petroleum-based fuels are decreasing day by day. Hence, many research workers are working hard to find an alternative fuel (Joshi et al. 2017). Due to around 20-fold increase in the consumption of fossil fuels in the past 100 years, there has been a large amount of depletion of fossil fuels due to which there has been adverse changes in climate globally. This is due to the large amount of greenhouse gas emissions produced by combustion of fossil fuels. Hence, many countries are giving high attention on use and production of alternate biofuels especially biodiesel. Countries like USA, Japan, China, and France are the leading producers of biodiesel. The biodiesel production in these countries has increased by over 2–32.3% since 2014. Figure 5.4 shows global biodiesel production between the years 2005–2012 (Hartikka et al. 2012).

Owing to their increased production of biodiesel, they have successfully implemented usage of biodiesel blends like B20 (20% biodiesel and 80% diesel). At present, about 3% of total fossil fuel usage has been substituted with biodiesel. The highest increase in biodiesel production globally was recorded in 2014 with an

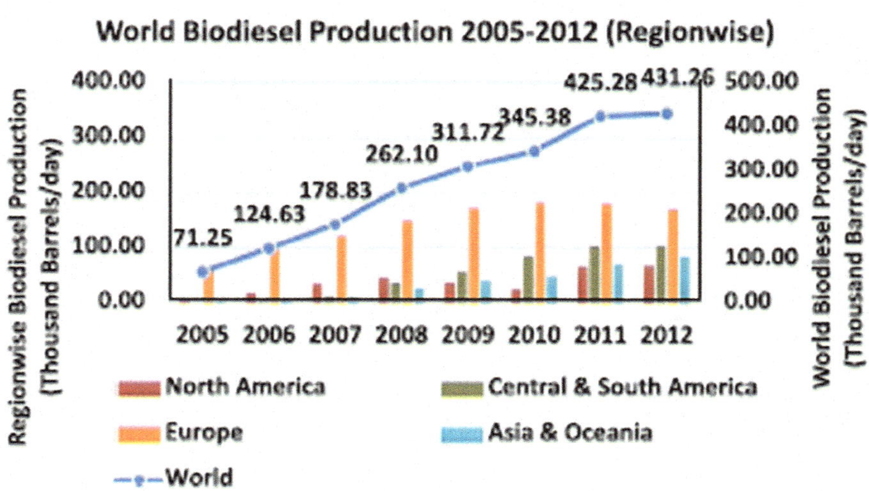

Fig. 5.4 Global biodiesel production between the years 2005–2012 (Joshi et al. 2017)

increase of 7.4%. According to current status, about 95% of the biodiesel produced worldwide is from edible oils like soybean, rapeseed. But in Brazil too along with soybean, animal fat is also used in large amount to produce biodiesel. About 94% of Brazil's biodiesel production is done using soybean and animal fats (Hartikka et al. 2012). Thailand took a major step toward reducing pollution by making it legal to use 7% palm methyl ester in commercially used diesel. Taking into consideration the heavy usage of edible oils for production of biodiesel leading to increasing food prices, Australia has shifted its 90% biodiesel production from edible oils to non-edible oils (Sugiyama et al. 2009; Zhang et al. 2009).

5.4 Sources and Methods of Production of Biofuels

Due to worthy fuel properties, high performance, and lower emissions, the demand for biodiesel has increased significantly. Hence, study of sources and techniques of production of biodiesel becomes a major topic of interest and study. Large number of vegetable oils and animal fat feedstocks can be used as source for biodiesel production. Earlier, vegetable oils were used directly in engines but later on this was found to cause many engine problems. Hence, these vegetable oils needed reduction of viscosity to be effectively used in engines (International Energy Outlook 2011). Two major methods of biodiesel formation are transesterification and esterification.

Since biodiesel production by transesterification involves the use of alcohol, this process is also called alcoholysis. Figure 5.5 shows the stepwise processes involved in transesterification method.

However in transesterification process, a lot of wastewater is generated after purification process and lot of time is consumed in pre-treatment process to reduce free fatty acid (FFA) content in the biodiesel produced. To avoid these drawbacks, solid acid catalyst derived from materials like coconut shell, sugar bagasse can be used to result in high biodiesel conversion of 94–95% (Escobar et al. 2009; Nigam and Singh 2011; Agarwal 2007; Demirbas 2007).

Biodiesel is also given another, (FAME) that is fatty acid methyl esters, since alcohols like methanol and ethanol are the most used alcohols for biodiesel production. 100% FAME is designated as B100. B20, B30, etc., are used to designate

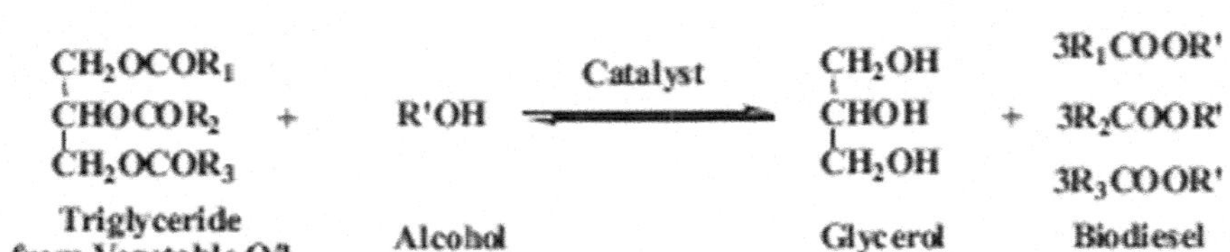

Fig. 5.5 Stepwise processes involved in transesterification method (Joshi et al. 2017)

biodiesel blends (Hossain and Davies 2010). Usually, various biofuels are classified into first generation, second generation, or third generation, etc., on the basis of materials that are used to produce that particular biofuel. Food crops that are edible feedstocks are the source of first generation of biofuels. Biodiesel produced by transesterification and ethanol produced by fermentation are all designated as first-generation biofuels. Edible oils included in the first-generation category are coconut raw oil, peanut raw oil, palm raw oil, sunflower raw oil. Biofuels which are produced by using non-edible feedstocks are referred to as second-generation biofuels. Edible oils included in this category are Pongamia raw oil, mahua raw oil, etc. (Murugesan et al. 2009; Graboski and McCormick 1998). Figure 5.6 shows various feedstocks used for biodiesel production.

Table 5.1 shows the various materials that are used to produce first-generation, second-generation, and third-generation biofuels. Table 5.2 makes a comparison of various advantages and disadvantages of first-generation and second-generation biofuels.

About 95% of total biodiesel production takes place from edible oils. Edible oils are more preferred for biodiesel production because of its low fatty acid content due to which biodiesel can be produced in single-step transesterification. Transesterification solves the drawbacks of first-generation raw materials like poor fuel atomization, engine deposits. Even though biofuels can be produced from both edible as well as

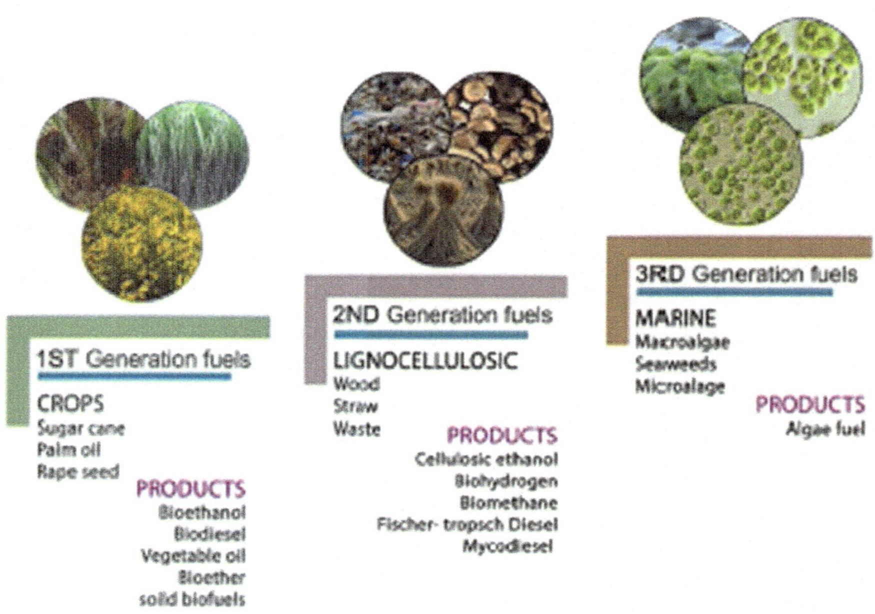

Fig. 5.6 Various feedstocks used for biodiesel production (Graboski and McCormick 1998)

Table 5.1 Different generations of biofuels (Kumar Vijay et al. 2017; Saravanan et al. 2018; Abdullah et al. 2017; Ahmad et al. 2011; Ashnani et al. 2014; Ayodele and Dawodu 2014; Babadi et al. 2016; Dawodu et al. 2014; Conceição et al. 2007; Dantas et al. 2007; Demirbas 2007, 2008; Beckman et al. 2018; Falade et al. 2008; Freire et al. 2009; Masjuki and Abul 2013; Knothe 2010; No 2014; Koyama et al. 2007; Joshi et al. 2017; Hartikka et al. 2012; Sugiyama et al. 2009; Zhang et al. 2009; International Energy Outlook 2011; Escobar et al. 2009; Nigam and Singh 2011; Agarwal 2007; Hossain and Davies 2010; Murugesan et al. 2009; Graboski and McCormick 1998; Lapuerta et al. 2008; Brito et al. 2012; Brunschwig et al. 2012; Rathore and Madras 2007; Borges and Díaz 2012; Bournay et al. 2005; Salamatinia et al. 2012; Dehkordi and Ghasemi 2012; Ilgen 2012; Noureddini et al. 2005; Marchetti et al. 2007; Jang et al. 2012; Van Kasteren and Nisworo 2007; Lee et al. 2011; Noureddini and Zhu 1997; Lotero et al. 2005; Mendonça et al. 2011; Charoenchaitrakool and Thienmethangkoon 2011; Silva et al. 2011; Zhou and Boocock 2006)

S. No.	Generation of biofuel	Materials used
1	First-generation biofuel	Soybean oil, corn grain, sugar beets, wheat, maize, forestry, sugarcane, cereals, rice, grain sorghum, sweet sorghum, cassava, barley, yams, oats, vegetable oil, animal fats, potatoes
2	Second-generation biofuel	Pongamia oil, jatropha oil, cassava, jackfruit seeds, aquatic weeds, cornfield wastes, mill wastes, flax, Brassica carinata crop, sawdust, corn stover barley straw, wheat straw, rice straw, sorghum straw, bagasse, cotton stalk, soft bamboo, rye, whey, tallow oil
3	Third-generation biofuel	Algae and microalgae

non-edible sources, non-edible sources are more preferred due to low cost of production as well as to reduce the use of edible food crops. Biodiesel is produced from edible oils by transesterification with solid acid catalysts so as to avoid saponification. Among the non-edible crops that can be used to produce biodiesel, Jatropha curcas has the best produce ability. Also low cost of biodiesel production from this crop makes it more favorable for large-scale production of biodiesel from this crop. The methods involved in biodiesel production from Jatropha curcas are plantation, oil extraction, transesterification, purifying, stabilizing, blending, and marketing. Non-edible oils are considered more favorable for biodiesel production because they can be cultivated on land which are not fit for cultivation of edible crops that too at low cost; they reduce CO_2 concentration in the region where their cultivation is done and because these oils are unfit for consumption, the use of these non-edible oils would not create scarcity of food. Another advantage of using non-edible oils for production of biodiesel is that if one raw material becomes scarce, the production can be fulfilled by using another raw material (Lapuerta et al. 2008; Brito et al. 2012; Brunschwig et al. 2012; Rathore and Madras 2007; Borges and Díaz 2012; Bournay et al. 2005; Salamatinia et al. 2012). Even though non-edible oil-derived biodiesels are good alternative fuels to fossil diesel as well as edible oil biodiesel, yet the production of biodiesel from non-edible oils is not economical and also the non-edible sources can have other uses like manufacturing of soap, medicine, etc. The limited availability

Table 5.2 Comparison of first and second generation of biofuels (Kumar Vijay et al. 2017; Saravanan et al. 2018; Abdullah et al. 2017; Ahmad et al. 2011; Ashnani et al. 2014; Ayodele and Dawodu 2014; Babadi et al. 2016; Dawodu et al. 2014; Conceição et al. 2007; Dantas et al. 2007; Demirbas 2007, 2008; Beckman et al. 2018; Falade et al. 2008; Freire et al. 2009; Masjuki and Abul 2013; Knothe 2010; No 2014; Koyama et al. 2007; Joshi et al. 2017; Hartikka et al. 2012; Sugiyama et al. 2009; Zhang et al. 2009; International Energy Outlook 2011; Escobar et al. 2009; Nigam and Singh 2011; Agarwal 2007; Hossain and Davies 2010; Murugesan et al. 2009; Graboski and McCormick 1998; Lapuerta et al. 2008; Brito et al. 2012; Brunschwig et al. 2012; Rathore and Madras 2007; Borges and Díaz 2012; Bournay et al. 2005; Salamatinia et al. 2012; Dehkordi and Ghasemi 2012; Ilgen 2012; Noureddini et al. 2005; Marchetti et al. 2007; Jang et al. 2012; Van Kasteren and Nisworo 2007; Lee et al. 2011; Noureddini and Zhu 1997; Lotero et al. 2005; Mendonça et al. 2011; Charoenchaitrakool and Thienmethangkoon 2011; Silva et al. 2011; Zhou and Boocock 2006)

Type	First-generation biofuels	Second-generation biofuels
	Edible oils	Non-edible oils
Feedstocks	Vegetable oils, corn, sugar, and animal fats	Aquatic biomass, forest residue, abundant plant waste biomass, and agricultural waste
Products	Biodiesel, corn ethanol, sugar alcohol	Biodiesel, butanol, mixed alcohols
Name of oils	Coconut, sunflower, palm, peanut, rapeseed, corn, groundnut oil, etc.	Jatropha, mahua, Pongamia, neem, rubber seed, castor oil, etc.
Problems	Restricted feedstock, mix together partly with diesel fuel	Available in forest areas
	Environmentally friendly for trade and industry, social safety	Cannot compete with foodstuff; It helps to progress to diminish the cost of conversion

of first and second generation of raw materials draws more interest toward biodiesel production from algal material. Because of the fact that large number of species of microalgae can be used to produce biodiesel immensely and microalgae can be grown and cultured on non-arable land, researchers have gathered interest on the study of biodiesel production using microalgae. Since a large number of steps are involved in conversion of microalgae into biodiesel, the production process from microalgae is not an economical option. Hence, research works are still continuing to find sustainable and economical ways of producing biodiesel from microalgae (Dehkordi and Ghasemi 2012).

Bio-lipids are also used widely for production of biodiesel. Among bio-lipids like vegetable oils, animal fats, waste cooking oils, vegetable oils are the most attractive option. Major oil-containing crops used for biodiesel production includes jatropha, Pongamia, soybean, sunflower, etc. Even though they can be used to successfully produce biodiesel and use them in engines, these oil crops have good market as cooking oils. Due to this, these oils are expensive and production of biodiesel from these is also expensive. Therefore, biodiesel production from non-edible oils is gaining more attention due to low expensiveness of production and good quality of biodiesel (Ilgen 2012). Crop species like buriti, murumuru, and pracaxi found in Brazil also have high potential to be used as source of biodiesel production. The biodiesel derived from these materials has good stability and high volatility which are necessary for good combustion, hence making them good enough to be used in diesel engines (Noureddini et al. 2005). Hydrogenated vegetable oil (HVO), a second-generation biodiesel source derived by catalytic hydrogenation of animals and waste cooking oil, has high potential to produce biodiesel to be used as engine fuel because of its high LHV, cetane number, low sulfur content, etc. Moringa oleifera, also known as drumstick tree, is a potential non-edible feedstock for production of biodiesel. This feedstock can also grow in areas of low rainfall and poor soil. It is majorly found in Africa, Asia, Latin America, and Oceana countries including Australia. Transesterification only is the most effective way to produce biodiesel from Moringa oleifera. Moringa seed oil is also considered as a prospective feedstock for biodiesel production. Moringa seed contains 38–40% oil from which bio diesel can be derived by catalytic or non-catalytic transesterification (Marchetti et al. 2007; Jang et al. 2012; Van Kasteren and Nisworo 2007).

5.5 Fuel Properties

Fuel properties are one of the most important characteristics that define the chemical and physical properties of that fuel. Knowing the properties of fuel is important for many applications like design of fuel systems in vehicles, dispensing system, storage system. Fuel properties of a fuel such as viscosity, density, cetane number, calorific value, flash point have a large impact on the performance and emission characteristics, efficiency and durability of engines whenever they are used in various types of engines (Lee et al. 2011). Cetane number increases the antiknock properties of the engine, high calorific value improves BSFC, high oxygen content improves combustion, and high flash point ensures low flammability of the fuel. Fuel properties also define how much safe or flammable, hazardous, explosive a fuel can be. Knowing the properties of fuel helps in molding the fuel in such a way that it can be used safely for running automobile vehicles and for other industrial applications. In the present situation, many types of additives like oxygenated additives, cetane number additives are used to increase the quality of fuel as well as performance of engine and reduce the emissions of engines (Noureddini and Zhu 1997). The fuel properties of diesel and biodiesel are almost comparable due to which biodiesel can be blended in any ratio with diesel and used in diesel engine without making any or very few changes in the engine. The diesel and biodiesel blend is usually represented by symbol BX where X denotes the percentage biodiesel in the blend. For example, B30 blend means 30% biodiesel is mixed with 70% diesel. As the percentage of biodiesel in the blend increases, performance parameters like BTE, BSFC increases. Best blend of biodiesel is always determined by evaluating its properties like density, viscosity, cetane number, flash point. Table 5.3 shows a comparison of the fuel properties of diesel and biodiesel (Lotero et al. 2005; Mendonça et al. 2011).

Due to high density, viscosity, and cetane number, the performance that biodiesel would give is similar to that of diesel. Since biodiesel can be blended in any ratio with diesel, it has proved to be a very good alternative substitute fuel. Due to higher flash point of biodiesel, the chances of fire are avoided due to low flammability. Due to high heating value and oxygen content and the absence of sulfur content, biodiesel also gives good combustion. Oxidation stability of biodiesel can be increased by using antioxidants like pyrogallol. But addition of antioxidants decreases the viscosity of biodiesel (Charoenchaitrakool and Thienmethangkoon 2011). Table 5.4 shows a comparison of properties of biodiesel produced in different countries. Due to the similarity in the properties of diesel and biodiesel, the performance given by diesel and biodiesel blends is also similar to the performance observed in case of diesel (Silva et al. 2011).

Due to the fact that various countries like USA, France use various materials for the production of biodiesel, they have developed their own standards of specification of biodiesel which are almost similar to one another (Saravanan et al. 2018). The storage, handling, and transportation of biodiesel are easy due to its high flash point (Zhou and Boocock 2006; Puhan et al. 2009; Radu et al. 2009; Nwafor et al. 2000; Caresana 2011; Purushothaman and Nagarajan 2009; Raheman and Ghadge 2008; Choi and Reitz 1999; Shivakumar et al. 2011; Huang et al. 2011; Qi et al. 2011).

There are various characteristics a fuel must satisfy so as to become of use in an engine. Various factors like combustion initiation time, fuel blending time depends on the fuel. To achieve high performance, a fuel must satisfy characteristics like high energy density, thermal stability, and less deposit-forming tendencies, toxicities, pollution. Biodiesel obtained from non-edible sources like mahua, jatropha, neem satisfies the above-mentioned characteristics and therefore are considered as good alternative fuel (Charoenchaitrakool and Thienmethangkoon 2011).

Table 5.3 Comparison of the fuel properties of diesel and biodiesel (Mendonça et al. 2011; Charoenchaitrakool and Thienmethangkoon 2011)

S. No.	Fuel properties	Diesel	Biodiesel
1	Density (kg/m^3)	847	876
2	Viscosity	1.3–4.2	2.0–6.1
3	Carbon (wt%)	0	12
4	Sulfur (wt%)	0.05	0
5	Hydrogen (wt%)	12	13
6	Flash point (°C)	61–80	100–169
7	Pour point (°C)	−34 to −15	−15 to 15
8	Cloud point (°C)	−15 to 5	−3 to 11
9	Boiling point (°C)	189–343	182–338
10	Cetane number	40–54	49–60

Table 5.4 Comparison of properties of biodiesel from different countries (Saravanan et al. 2018; Mendonça et al. 2011; Silva et al. 2011)

S. No.	Country	Specification name	Density (kg/m^3)	Viscosity	Flash point	Cetane number
1	India	BIS-15607	870–890	1.9–6	130	40
2	Germany	DIN-EN-590	875–890	3.5–5	110	49
3	USA	ASTM-424720	880	1.96	130	47
4	Australia	ONC-1191	850–890	3.5–5	100	49
5	France	EU-15412	870–890	3.5–5	100	49
6	Italy	UNI-10946	860–890	3.5–5	100	–

5.6 Engines Used for Testing

The performance parameters like BTE, BSFC and emissions parameters like HC, CO, No_x emissions depend not only on the fuel properties but also on the type of engine in which the fuel is tested (Abdullah et al. 2017; Ahmad et al. 2011). The speed, power, compression ratio, number of cylinders, etc., play a major role in determining the performance and emission characteristics.

5.6.1 Performance of Engines Fueled with Biodiesel Fuel and Its Various Blends

Table 5.5 shows a summary of the effects on the performance of various engines when they were fueled with biodiesel and its various blends. The results are based on various performance parameters like brake thermal efficiency (BTE), brake-specific fuel consumption (BSFC), brake mean effective pressure (BMEP), brake horsepower (BHP), mechanical efficiency (Cenk and Metin 2011; Ganapathy et al. 2011; Kannan and Anand 2011; Pandian et al. 2011; Jaichandar et al. 2012; Kuti et al. 2013). In diesel and biodiesel blends, it has been found that with increase in percentage of biodiesel in the blends, BSFC and BTE increases.

When various types of additives like oxygenated additives, antioxidant additives, metal-based additives were used along with biodiesel fuel, it showed considerable improvement in the performance of the diesel engine. The most useful trend is that BSFC increases with increase in biodiesel quantity in the fuel blend. Many researchers investigated on the effect of additives on the performance of engines. Puhan et al. (2009) found through experiments that mahua methyl ester and mahua ethyl ester are good substitutes to be used in CI engines in place of diesel. When various biodiesel blends like jatropha (B100), canola oil (B20), tall oil (B60), palm oil (B20) was mixed with additives like (Al–Mg), Mn, Ni, BHT, TBHQ, then it was found that there is a decrease in BSFC by about 2.1–10%.

Table 5.5 Effects on the performance of various engines when they were fueled with biodiesel and its various blends

Fuel	Engine	Findings	References
Jatropha oil (B100) + Al–Mg	Variable load constant speed diesel engine	BSFC decreased with 3% addition of Al–Mg and BTE increased with 1% addition of Al–Mg	Ayodele and Dawodu (2014)
Jatropha oil (B15) + Cerium oxide nanoparticles	Variable load constant speed diesel engine	BSFC decreased by 3% and BTE increased by 1.7%	Dawodu et al. (2014)
Tall oil (B60) + Mn	Variable load constant speed diesel engine	BSFC decreased with 10% addition of Mn	Conceição et al. (2007)
Tall oil (B60) + Ni	Variable load constant speed diesel engine	BSFC decreased with 5% addition of Ni	Dantas et al. (2007)
Canola oil (B20) + BHA	Variable load constant speed diesel engine	BSFC decreased by 4.09%	Demirbas (2008)
Canola oil (B20) + EHN	Variable load constant speed diesel engine	BSFC decreased by 4.12%	Freire et al. (2009)
Canola oil (B20) + EHN	Variable load constant speed diesel engine	BSFC decreased by 4.22%	Joshi et al. (2017)
Canola oil (B20) + TBHQ	Variable load constant speed diesel engine	BSFC decreased by 10.19%	Sugiyama et al. (2009)
Palm oil (B20) + BHA	Variable load constant speed diesel engine	BSFC decreased by 0.44%. BTE increased by 0.92%	Escobar et al. (2009)
Palm oil (B20) + BHT	Variable load constant speed diesel engine	BSFC decreased by 0.68%. BTE increased by 0.37%	Murugesan et al. (2009)
B100 + D-16 Mg	Variable load constant speed diesel engine	BSFC decreased by 2.01%	Rathore and Madras (2007)
Plastic oil biodiesel	Variable load constant speed diesel engine	BSFC decreased, whereas BTE increased	Borges and Díaz (2012)
B100 + D-8 Mg	Variable load constant speed diesel engine	BSFC decreased by 1.02%	Jang et al. (2012)
Ethylic biodiesel fuel	CI engine	BSFC increased by 9%	Marchetti et al. (2007)

(continued)

Table 5.5 (continued)

Fuel	Engine	Findings	References
Rapeseed oil biodiesel (ROB)	CI engine	BSFC increased by 3–14%	Knothe (2010)
Waste oil biodiesel	CI engine	BSFC increased by 20%	Mendonça et al. (2011)
Palm oil biodiesel	CI engine	BSFC increased by 20%	Charoenchaitrakool and Thienmethangkoon (2011)
B20 biodiesel	CI engine	BSFC increased by 20.13% and efficiency increased by 3%	Caresana (2011)
B20 biodiesel	4276T Turbo-Charged Diesel engine	Increase in efficiency and fuel consumption	Purushothaman and Nagarajan (2009)
Biodiesel + diesel + 10% methanol + 10% fumigation methanol	Four-cylinder direct injection at constant speed (1800 r.p.m) engine	BSFC and BTE increased at high loads	Cenk and Metin (2011)
B5 and B20 blends of biodiesel + diesel	Diesel engine	BSFC increased by 11%	No (2014)
Anhydrous ethanol diesel fuel Biodiesel (B-100) Natural gas	IC engine	Efficiency increased by about 77.34–91.15%	Sugiyama et al. (2009)
B100 + diesel	Four cylinder turbo-charged	BSFC increased by 13.8%	Qi et al. (2011)
B100 + diesel	Diesel engine	BSFC increased by 11.4%	Silva et al. (2011)
Sunflower oil biodiesel + diesel	Single-cylinder diesel engine	BTE, BP and BMEP decreased, whereas BSFC increased slightly	Noureddini and Zhu (1997)
Kernel oil biodiesel and diesel oil	Single-cylinder diesel engine	Combustion improved significantly	Kannan and Anand (2011)
Ethanol + biodiesel + diesel	Single-cylinder diesel engine	BSFC and BTE increased significantly	Raheman and Ghadge (2008)
B100 + diesel	Four-cylinder diesel engine	BTE decreased by 32.4%, whereas BSFC increased	Puhan et al. (2009)
iso-HVO + biodiesel	Diesel engine	BSFC and BTE increased significantly	Radu et al. (2009)

(continued)

Table 5.5 (continued)

Fuel	Engine	Findings	References
GTL + HVO	DI engine	Ignition delay was shortened and fuel consumption was less	Purushothaman and Nagarajan (2009)
SME	Single-cylinder, direct injection diesel engine	BSFC increased by 14.65%, whereas BTE decreased by 8.07%	Raheman and Ghadge (2008)
20% SME	Single-cylinder, direct injection diesel engine	BSFC increased by 4.2%, whereas BTE decreased by 2.61%	Demirbas (2007)
40% SME	Single-cylinder, direct injection diesel engine	BSFC increased by 8.7%, whereas BTE decreased by 4.95%	Hossain and Davies (2010)
B20 biodiesel blend	CI engine	BTE increased by 1.61–10.79%. BSFC increased by 0.295–2.96%	Murugesan et al. (2009)
B20A biodiesel blend	CI engine	BTE increased by 0.12–6.05%. BSFC increased by 0.0295–1.91%	Ilgen (2012)
B10, B25 Moringa seed oil biodiesel	Single-cylinder, four-stroke diesel engine	BSFC increased slightly, whereas BTE decreased slightly	Lee et al. (2011)
B10, B20 Moringa seed oil biodiesel	Four-Cylinder, four-stroke diesel engine	BSFC increased by 5.13–8.39%. BTE and BP decreased	Raheman and Ghadge (2008)
B0, B10 Moringa seed oil biodiesel	Four-cylinder, four-stroke diesel engine	BSFC increased slightly, whereas BP decreased by 4%	Choi and Reitz (1999)
B5, B10 Moringa seed oil biodiesel	Four-cylinder, four-stroke diesel engine	BSFC increased by 5%. BTE and BP decreased	Shivakumar et al. (2011)

The observation made in case of BTE was that it increases whenever biodiesel blends were mixed with various additives and the increase was about 0.37–1.7% (Radu et al. 2009). Engine performance, especially cold-weather performance of biodiesel, is less as compared to diesel fuel. This drawback can be overcome by using flow-improving additives and by maintaining proper engine timing. Even though air pollution is very less when biodiesel is used, clogging and deposit problems do occur frequently. This might be because of either lower quality of biodiesel used or due to low stability of oxidation (Nwafor et al. 2000). Of all the biodiesel and diesel fuel blends, B20 blend is proven to be more efficient as its high lubricity causes less wear and tear to engine parts. In CI engines, B100 biodiesel has been found to increase performance and reduce emission to a large extent. When biodiesel derived from various sources like waste cooking oil, palm oil were tested on CI engine, it was found that BSFC increased by about 3–20% accompanied with increase in efficiency. When diesel and biodiesel blends were used in various modified engines like 4276T Turbo-Charged Diesel engine, four-cylinder diesel engine, it was found that there was considerable increase in BSFC by about 11–13.8% which is more than the BSFC improvement found in case of diesel fuel. When various biodiesel blends were used in modified engines, it was found that BSFC, BHP, and BMEP increased, whereas BTE reduced maximum times. Due to the compatibility of biodiesel with mineral diesel, biodiesel can be used as alternate fuel for mineral biodiesel and it has been proven that engine performance is also similar with no or very few modifications in the engine (Caresana 2011; Purushothaman and Nagarajan 2009; Raheman and Ghadge 2008; Choi and Reitz 1999). When single-cylinder, direct injection diesel engine was fueled with soybean methyl ester (SME) and its 20 and 40% blends with diesel fuel, it was found that BSFC increased by 4.2–14.65% with increase of SME in the blend from, whereas BTE decreased by 2.61–8.07% with increase of SME in the blend. Antioxidants are added in along with biodiesel in order to increase oxidation stability. But even though there is increase in performance of engine when fueled with fuel mixed with antioxidant, the increase is less in comparison to biodiesel blend without antioxidant (Shivakumar et al. 2011).

5.6.2 Emissions of Engines Fueled with Biodiesel Fuel and Its Blends

Table 5.6 shows a summary of the effects on the emissions produced by various engines when they were fueled with biodiesel and its various blends. The results are based on the amount of HC, CO, CO_2, PM, smoke, and NO_x emissions produced by engines when they are run on biodiesel fuel blends.

HC emissions usually decrease with the use of biodiesel fuel and its blends due to the presence of oxygen as well as high cetane number of biodiesel. Due to these factors, the fuel undergoes complete combustion and ignition delay is reduced (Huang et al. 2011). Whenever biodiesel fuel blends are mixed with additives like Al–Mg, cobalt oxide, BHA, BHT, the HC emissions reduced by about 23.81–76%. Hence, cetane number additives proved to be most effective in reducing HC emissions. CO emissions are toxic, colorless, and odorless emissions which are harmful for the environment. When biodiesel fuel blends are used in various engines, CO emissions usually reduce due to high oxygen content and cetane number of biodiesel. When investigated with additives like Al–Mg, EHN, DEE, it was found that the CO emissions reduced by about 13.43–66%. Among the major

Table 5.6 Effects on the emissions produced by various engines when they were fueled with biodiesel and its various blends

Fuel	Engine	Emissions results	References
Jatropha (B100) + Al–Mg	Variable load constant speed diesel engine	HC reduced by 76%. CO reduced by 66%	Ayodele and Dawodu (2014)
Jatropha (B100) + cobalt oxide	Variable load constant speed diesel engine	HC reduced by 52%. CO reduced by 33%	Dawodu et al. (2014)
Jatropha (B100) + cerium oxide	Variable load constant speed diesel engine	NO_x emissions reduced by 23.5%	Conceição et al. (2007)
Mahua biodiesel + 5% diethyl ether	Variable load constant speed diesel engine	HC decreased	Dantas et al. (2007)
Biodiesel + BHA + BHT	Variable load constant speed diesel engine	HC reduced by 23.81%	Demirbas (2008)
Mahua Oil + 15% DEE	Variable load constant speed diesel engine	CO reduced by 67%. NO_x reduced at low loads	Freire et al. (2009)
Palm oil (B20) + BHA	Variable load constant speed diesel engine	CO reduced by 21.21%. NO_x reduced drastically	Joshi et al. (2017)
Palm oil (B20) + BHT	Variable load constant speed diesel engine	CO reduced by 23.81%. NO_x reduced drastically	Sugiyama et al. (2009)
Biodiesel + Mg	Variable load constant speed diesel engine	CO reduced by 13.43%	Escobar et al. (2009)
Biodiesel + Mn	Variable load constant speed diesel engine	CO reduced by 16.35%	Murugesan et al. (2009)
Plastic oil biodiesel + DEE	Variable load constant speed diesel engine	CO reduced drastically with complete combustion. NO_x decreased by 29%	Rathore and Madras (2007)
Mahua oil (B100) + DMC	Variable load constant speed diesel engine	CO reduced drastically with better air–fuel ratio	Borges and Díaz (2012)
Cottonseed biodiesel + 0.01% L-ascorbic acid	Variable load constant speed diesel engine	NO_x decreased by 5.65%. CO2 decreased by 1%	Jang et al. (2012)

(continued)

Table 5.6 (continued)

Fuel	Engine	Emissions results	References
Fish biodiesel + 2% DEE	Variable load constant speed diesel engine	CO_2 decreased drastically	Marchetti et al. (2007)
Ethylic biodiesel fuel	CI engine	CO emission reduced by 14%	Knothe (2010)
Rapeseed oil biodiesel (ROB)	CI engine	NO_x increased by 8%	Charoenchaitrakool and Thienmethangkoon (Mendonça et al. 2011)
Waste oil biodiesel	CI engine	NO_x increased by 20%	Charoenchaitrakool and Thienmethangkoon (2011)
Palm oil biodiesel	CI engine	HC reduced by 22%	Caresana (2011)
B20 biodiesel	CI engine	NO_x increased by 11.2%	Purushothaman and Nagarajan (2009)
B100 + diesel	High-speed direct injection engine	CO and HC emissions reduced drastically	Cenk and Metin (2011)
Cottonseed oil biodiesel	Diesel engine	CO emission and PM reduced, whereas NO_x emissions increased slightly	No (2014)
B100 + diesel	Marine craft engine	CO emissions reduced drastically	Sugiyama et al. (2009)
Soybean and yellow grease biodiesel + ULSD	Single-cylinder Horizontal type	CO and HC emissions reduced with low soot formation. NO_x emissions increased slightly	Qi et al. (2011)
Biodiesel + diesel + 10% methanol + 10% fumigation methanol	Four-cylinder direct injection at constant speed (1800 r.p.m) engine	Considerable reduction in CO_2, NO_x and soot formations	Silva et al. (2011)
B5 and B20 blends of biodiesel + diesel	Diesel engine	Smoke emissions reduced by 60%	Noureddini and Zhu (1997)
B10 biodiesel blend	Common rail engine	PM emission reduced, whereas NO_x emissions increased marginally	Kannan and Anand (2011)
B100 + diesel	Single-cylinder engine with exhaust gas recirculation	PM and CO emissions reduced, whereas NO_x emissions increased	Raheman and Ghadge (2008)

(continued)

Table 5.6 (continued)

Fuel	Engine	Emissions results	References
B100 + diesel	Four-cylinder turbo-charged (1400 rpm)	CO and HC emissions reduced significantly. NO_x emission increased by 11.2%	Puhan et al. (2009)
B100 + diesel	Diesel engine	CO emission decreased by 12%, whereas NO_x emission decreased by 20%	Radu et al. (2009)
Soybean biodiesel	Six-cylinder diesel engine	Increase in NO_x emission due to higher consumption of biodiesel	Purushothaman and Nagarajan (2009)
Karanja biodiesel + diesel	Four-cylinder turbo-charged direct injection diesel engine	NO_x and smoke emissions reduced significantly	Raheman and Ghadge (2008)
B100 + diesel	Four-cylinder CI engine	Soot emissions reduced significantly	Demirbas (2007)
Soybean biodiesel + diesel	Four stroke direct injection diesel engine	Soot emission reduced, whereas NO_x emission reduced slightly	Hossain and Davies (2010)
B100 + diesel	Four-cylinder diesel engine	CO emission reduced by 27.7%	Murugesan et al. (2009)
Rapeseed oil biodiesel	Petter model AC1 diesel engine	CO emissions reduced at advanced IT. HC reduced by 13% at advanced IT and increased at retarded IT. CO_2 emissions reduced by 4.4% at advanced IT	Ilgen (2012)
Soybean biodiesel	Caterpillar 1Y3700 SCOTE D	CO emissions increased at low loads and poor correlation between CO and PM emissions was observed at high loads. HC emissions increased at advanced IT and reduced at retarded IT. NO_x increased at advanced IT and reduced at retarded IT. PM emissions reduced for all variations of IT	Lee et al. (2011)
B5 biodiesel blend	Lombardini 6 LD 400 engine	CO emission decreased by 13.54% at advanced IT and increased by 1.21% at retarded IT. Smoke emissions reduced by 1.25% at advanced IT and increased by 1.02% at retarded IT	Raheman and Ghadge (2008)

(continued)

Table 5.6 (continued)

Fuel	Engine	Emissions results	References
Jatropha biodiesel	Greaves Cotton model GL 400 II A engine	CO emission increased slightly at advanced IT and reduced at retarded IT. HC emission reduced by 13.54% at advanced IT and increased by 1.34% at retarded IT. NO_x increased at advanced IT and reduced at retarded IT. Smoke emissions reduced at advanced IT and vice versa	Choi and Reitz (1999)
Waste cooking oil methyl ester	TV1-KIRLOSKAR engine	CO emission reduced by 0.2% at advanced IT. HC emissions increased at advanced IT and reduced at retarded IT. NO_x increased at advanced IT and reduced by 19% at retarded IT. Smoke emission reduced at advanced IT and vice versa. CO_2 emissions increases for all IT's	Shivakumar et al. (2011)
Pongamia biodiesel	KIRLOSKAR engine	CO emissions reduced at reduced at advanced IT and increased at retarded IT. HC emissions reduced at advanced IT and increased at retarded IT. PM reduced at advanced IT. Smoke emissions reduced at advanced IT and vice versa	Ayodele and Dawodu (2014)
Pongamia biodiesel	TV1-KIRLOSKAR engine	CO emissions reduced at reduced at advanced IT and increased at retarded IT. HC emissions reduced at advanced IT and increased at retarded IT. NO_x increased at advanced IT and reduced at retarded IT. Smoke emissions reduced at advanced IT and vice versa	Dawodu et al. (2014)
Mahua biodiesel	TV1-KIRLOSKAR engine	CO emissions reduced at advanced IT and increased at retarded IT. NO_x emission increased by 39% for advanced IT and reduced by 11.8–28% at retarded IT	Conceição et al. (2007)
Soybean biodiesel	KIRLOSKAR engine	CO emissions reduced at retarded IT and increased at advanced IT	Dantas et al. (2007)
Waste cooking oil biodiesel	Daedong ND 130DI engine	CO emissions reduced at retarded IT and increased at advanced IT. No change in HC emissions at advanced IT and increased by 34% at retarded IT. Smoke emissions increased at advanced IT and vice versa	Demirbas (2008)

(continued)

Table 5.6 (continued)

Fuel	Engine	Emissions results	References
B100 biodiesel blend	Multi-cylinder turbo-charged-type engine	CO emissions increased by 22.3% at retarded IT. HC emissions reduced by 13% at advanced IT and increased for retarded IT. NO_x emissions increased by 5–12.34% at advanced IT and decreased by 7–67% for retarded IT	Freire et al. (2009)
80% Jatropha + 20% tyre pyrolysis oil	Kirloskar TAF1	CO emissions reduced at advanced IT by 13.3% and increased at retarded IT by 16.8%. HC emissions reduced by 14.2% at advanced IT and increased by 14.6% at retarded IT. NO_x emissions increased by 4.9% for advanced IT and reduced by 14.9% for retarded IT. PM increased by 9.3% at all IT's	Joshi et al. (2017)
Karanja biodiesel	Kirloskar TAF1	CO emissions reduced at advanced IT and increased at retarded IT. HC emissions reduced at advanced IT and increased at retarded IT. NO_x increased at advanced IT and reduced at retarded IT. PM reduced at advanced IT	Sugiyama et al. (2009)
Annona biodiesel	KIRLOSKAR engine	CO emissions reduced at advanced IT and increased at retarded IT. HC emissions reduced at advanced IT and increased at retarded IT. Smoke emissions increased at advanced IT and vice versa	Escobar et al. (2009)
Rice bran oil biodiesel	Diesel engine	NO_x increased at advanced IT and reduced at retarded IT. Smoke emissions reduced at advanced IT and vice versa	Murugesan et al. (2009)
HVO	Diesel engine	PM, CO and HC emissions reduced by 50%. NO_x emission was similar to that of diesel fuel	Rathore and Madras (2007)
iso-HVO + biodiesel	Diesel engine	HC and CO emissions reduced, whereas PM and NO_x emissions were similar to that of diesel fuel	Borges and Díaz (2012)
30% HVO + biodiesel	Diesel engine	NO_x emissions increased	Jang et al. (2012)
GTL + HVO	DI engine	HC, CO and PM emissions reduced, whereas NO_x emission increased	Marchetti et al. (2007)

(continued)

Table 5.6 (continued)

Fuel	Engine	Emissions results	References
Hydro-processed renewable diesel (HRD)	Diesel engine	HC, CO and PM emissions reduced significantly, whereas NO_x emissions increased which was still less than 29% in comparison to diesel fuel	Knothe (2010)
SME	Single-cylinder, direct injection diesel engine	HC emission reduced by 38.4%. CO emission reduced by 41.7%. NO_x emission increased. Smoke emission reduced by 48.23%	Mendonça et al. (2011)
20% SME	Single-cylinder, direct injection diesel engine	HC emission reduced by 15%. CO emission reduced by 11.37%. NO_x emission increased by 7.5%. Smoke emission reduced by 20.5%	Charoenchaitrakool and Thienmethangkoon (2011)
40% SME	Single-cylinder, direct injection diesel engine	HC emission reduced by 27%. CO emission reduced by 29%. NO_x emission increased. Smoke emission reduced by 33.41%	Carcsana (2011)
B20 biodiesel blend	CI engine	NO_x emission increased by 1.24–5.33%. CO_2 emission increased by 3.19–5.49%. CO emission reduced by 16.67–33.33%. HC emission reduced by 2.33–6.98%	Purushothaman and Nagarajan (2009)
B20A biodiesel blend	CI engine	NO_x emission increased by 0.31–2.8%. CO_2 emission increased by 2.13–3.64%. CO emission reduced by 24.42–29.07%. HC emission reduced by 2.86%	Cenk and Metin (2011)
B10, B25 Moringa seed oil biodiesel	Single-cylinder, four-stroke diesel engine	PM, HC and CO emissions decreased, whereas NO_x emission increased	No (2014)
B10, B20 Moringa seed oil biodiesel	Four-cylinder, four-stroke diesel engine	HC and CO emissions decreased, whereas NO_x emission increased	Sugiyama et al. (2009)
B0, B10 Moringa seed oil biodiesel	Four-cylinder, four-stroke diesel engine	HC and CO emissions decreased, whereas NO_x emission increased	Qi et al. (2011)
B5, B10 Moringa seed oil biodiesel	Four-cylinder, four-stroke diesel engine	HC and CO emissions decreased	Silva et al. (2011)

gases emitted by diesel engines, NO_x is the most dangerous and hence researches have worked a lot on reducing NO_x emissions. On adding additives like BHA, BHT, DEE with biodiesel, there was significant decrease in NO_x emissions by about 5.65–29%. Use of biodiesel as a fuel increases CO_2 emissions due to high level of oxygen and low level of carbon in biodiesel. However, this problem can be overcome by using several additives which can reduce CO_2 emissions by about 1–2%. When biodiesel derived from various sources like waste cooking oil, palm oil were tested on CI engine, it was found that HC and CO emissions reduced by about 14–22%, whereas there was an increase in NO_x emission by 8–11.2% (Qi et al. 2011). When diesel and biodiesel blends were used in various modified engines like 4276T Turbo-Charged Diesel engine, four-cylinder diesel engine, it was found that CO and smoke emissions reduced considerably but NO_x emissions were on the higher side in comparison to diesel fuel. Among all the biodiesel and diesel blends, B20 and B100 are the most used and studied variants.

NO_x emissions can be reduced by adding additives, varying injection timings, or using catalytic converters. Still B20 biodiesel blend is more preferred than B100 because even though emissions decrease with increase in biodiesel, performance parameters decrease with increase in quantity of biodiesel mixed. When various biodiesel blends were used in modified engines, the reduction of HC, CO, and CO_2 emissions was similar to that of normal diesel engine but there was significant decrease in NO_x emissions. Use of biodiesel can reduce emission of polycyclic aromatic hydrocarbons by 75–90% and unburnt hydrocarbons by 90%. Varying injection timings has been found to be an effective way to reduce various emissions (Pandian et al. 2011). When various modified engines like TV1-KIRLOSKAR engine, Greaves Cotton model GL 400 II A engine were fueled with biodiesel obtained from various sources like Pongamia, jatropha, waste cooking oil, rice bran oil, it was found that at advanced IT's CO emissions reduced by 0.2–13.54%, HC emissions reduced by 13–14.2%, NO_x emissions increased by 5–12.34%, and CO_2, PM, and smoke showed slight decrement, whereas at retarded IT's, CO emission increased by 1.25–22.4%, HC emission increased by 1.34–34%, NO_x emission reduced by 7–67%, and CO_2, PM, and smoke emissions showed almost no change or increased slightly. When diesel engine was tested with hydrogenated vegetable oil (HVO) and its 20, 50, and 80% by mass fraction of hydrogenated vegetable oil blended with B7 commercial biodiesel, it was found that CO, HC, PM, and smoke emissions reduced significantly, whereas NO_x emission increased but this increase was about 29% less than the increase in emission observed when the engine was fueled with diesel. When single-cylinder, direct injection diesel engine was fueled with soybean methyl ester (SME) and its 20 and 40% blends with diesel fuel, it was found that with increase in SME in the blend, HC emissions reduced 15–38.4%, CO emissions reduced by 11.37–41%, smoke emissions reduced by 20.5–48.3% but NO_x emissions increased by 7–9%. Addition of antioxidants to diesel–biodiesel blends helped in reducing HC, CO, NO_x, and CO_2 emissions in comparison to pure diesel–biodiesel blend (Jaichandar et al. 2012; Kuti et al. 2013).

5.7 Major Findings

(1) In order to meet growing demand of fuel and to reduce foreign import of fossil fuels, India has implemented a national biodiesel mission which aims at replacing 20% of fossil fuel usage by biodiesel, i.e., to make use of B20 blend. For this, biodiesel has been produced from Karanja, Jatropha curcas, etc., in various parts of the nation.
(2) Non-edible oils are more preferred than edible oil due to low cost of production, to avoid scarcity of food that might occur due to excess use of edible oils for biodiesel production, and also because non-edible crops can be grown on non-arable land. Jatropha is the most preferred among non-edible raw materials.
(3) Fuel properties are one of the most important characteristics that define the chemical and physical properties of that fuel. Cetane number increases the antiknock properties of the engine, high calorific value improves BSFC, high oxygen content improves combustion, and high flash point ensures low flammability of the fuel. Fuel properties also define how much safe or flammable, hazardous, explosive a fuel can be. Knowing the properties of fuel helps in molding the fuel in such a way that it can be used safely for running automobile vehicles and for other industrial applications.
(4) Due to the highly oxygenated nature of biodiesel, it gives better engine performance and low harmful emissions. Due to high density, viscosity, and cetane number, the performance that biodiesel would give is similar to that of diesel. Since biodiesel can be blended in any ratio with diesel, it has proved to be a very good alternative substitute fuel. Due to higher flash point of biodiesel, the chances of fire are avoided due to low flammability. Due to high heating value and oxygen content and the absence of sulfur content, biodiesel also gives good combustion.
(5) Various performance parameters like brake thermal efficiency (BTE), brake-specific fuel consumption (BSFC), brake mean effective pressure (BMEP), brake horsepower (BHP), mechanical efficiency determine the performance of an engine when run on a particular fuel. In diesel and biodiesel blends, it has been found that with increase in percentage of biodiesel in the blends, BSFC and BTE increases.
(6) When various types of additives like oxygenated additives, antioxidant additives, metal-based additives were used along with biodiesel fuel, it showed considerable improvement in the performance of the diesel engine. The most useful trend is that BSFC increased by 2.1–10%. With increase in biodiesel quantity in the fuel blend, whereas BTE increased by 0.3–0.7%.
(7) When CI and other modified diesel engines were fueled with B20 and B100 biodiesel blends, it was found that BSFC increased by 3–20% along with increase in BHP, BMEP but BTE showed a decreasing trend. Soybean methyl ester (SME)-blended biodiesel showed increase in BSFC, whereas BTE decreased.

(8) The emissions produced by various engines when they were fueled with biodiesel and its various blends are HC, CO, CO_2, PM, smoke, and NO_x emissions. When CI and other modified diesel engines were fueled with B20 and B100 biodiesel blends, it was found that HC emission decreased by 20–67%, CO emission decreased by 12–48%, PM reduced by 2–47%, whereas NO_x emission increased by 2%.

5.8 Conclusions

Due to its good antiknock qualities, renewability, sustainability, compatibility with various types of engines and ability to increase engine performance and reduce emissions, biodiesel has been found to be an effective alternative fuel. Better performance of engines can be achieved by using biodiesel fuel along with additives like metal-based additives, cetane number additives, antioxidant additives. This will help in complete fuel combustion and increased engine performance. Whenever engines are fueled with biodiesel fuel, emissions can be further reduced by varying parameters like injection timing, injection pressure. In order to reduce the dependency on foreign imports of fuels, to meet growing demand for fuel, and to reduce harmful emissions, major energy-consuming countries have implemented production and use of biodiesel fuel. Noticeable among them is India. India has implemented national biodiesel mission which aims at replacing 20% of fossil fuel usage by biodiesel, i.e., to use B20 blends. For this, India has been producing biodiesel in various parts within the nation using crops like jatropha and Karanja. These initiatives will surely help in meeting increasing demand for fuel, and at present, 3% of total fossil fuel usage has been substituted by biodiesel. Even though biodiesel can be produced from edible, non-edible, and algal material, non-edible materials are more preferred for biodiesel production because of low production cost, high yield, and ability to be cultivated on non-arable land. Most preferred among non-edible sources are jatropha and Karanja. The most effective method for production of biodiesel is transesterification. Biodiesel has high density, viscosity, cetane number, and flash point due to which it can give good engine performance similar to that of diesel fuel. Use of biodiesel increases performance parameters like BSFC, BTE, BHP and BMEP to a large extent, whereas in case of emissions, HC, CO, CO_2, PM, and smoke emissions reduce but NO_x emissions have been found to be on the higher side.

Acknowledgements Authors thank VIT University, Vellore, and Amity University, Noida, Uttar Pradesh, for their support in this research work.

References

Abdullah SHYS, Hanapi NHM, Azid A, Umar R, Juahir H, Khatoon H, Endut A (2017) A review of biomass-derived heterogeneous catalyst for a sustainable biodiesel production. Renew Sustain Energy Rev 70:1040–1051

Agarwal AK (2007) Biofuels (alcohols and biodiesel) applications as fuels for internal combustion engines. Prog Energy Combust Sci 33:233–271

Ahmad AL, Mat Yasin NH, Derek CJC, Lim JK (2011) Microalgae as a sustainable energy source for biodiesel production: a review. Renew Sustain Energy Rev 15:584–593

Ashnani MHM, Johari A, Hashim H, Hasani E (2014) A source of renewable energy in Malaysia, why biodiesel? Renew Sustain Energy Rev 35:244–257

Ayodele OO, Dawodu FA (2014) Production of biodiesel from Calophyllum inophyllum oil using a cellulose-derived catalyst. Biomass Bioenergy 70:239–248

Babadi FE, Hosseini S, Soltani SM, Aroua MK, Shamiri A, Samadi M (2016) Sulfonated beet pulp as solid acid catalyst in one-step esterification of industrial palm fatty acid distillate. J Am Oil Chem Soc 93:319–327

Beckman J, Gooch E, Gopinath M, Landes M (2018) Market impacts of China and India meeting biofuel targets using traditional feedstocks. Biomass Bioenergy 108:258–264

Borges ME, Díaz L (2012) Recent development on heterogeneous catalysts for biodiesel production by oil esterification and transesterification reactions: a review. Renew Sustain Energy Rev 16:2839–2849

Bournay L, Casanave D, Delfort B, Hillion G, Chodorge JA (2005) New heterogeneous process for biodiesel production: a way to improve the quality and the value of the crude glycerin produced by biodiesel plants. Catal Today 106:190–192

Brito JQA, Silva CS, Almeida JS, Korn MGA, Korn M, Teixeira LSG (2012) Ultrasound-assisted synthesis of ethyl esters from soybean oil via homogeneous catalysis. Fuel Process Technol 95:33–36

Brunschwig C, Moussavou W, Blin J (2012) Use of bioethanol for biodiesel production. Prog Energy Combust Sci 38:283–301

Caresana F (2011) Impact of biodiesel bulk modulus on injection pressure and injection timing. The effect of residual pressure. Fuel 90:477–485

Cenk S, Metin G (2011) Impact of compression ratio and injection parameters on the performance and emissions of a DI diesel engine fueled with biodiesel-blended diesel fuel. Appl Therm Eng 31:3182–3188

Charoenchaitrakool M, Thienmethangkoon J (2011) Statistical optimization for biodiesel production from waste frying oil through two-step catalyzed process. Fuel Process Technol 92:112–118

Choi CY, Reitz RD (1999) An experimental study on the effects of oxygenated fuel blends and multiple injection strategies on DI diesel engine emissions. Fuel 78:1303–1317

Conceição MM, Candeia RA, Silva FC, Bezerra AF, Fernandes VJ Jr, Souza AG (2007) Thermoanalytical characterization of castor oil biodiesel. Renew Sustain Energy Rev 11: 964–975

Dantas MB, Conceição MM, Fernandes VJ Jr, Santos N, Rosenhaim R, Marques ALB, Santo IMG, Souza AG (2007) Thermal and kinetic study of corn biodiesel obtained by the methanol and ethanol routes. J Therm Anal Calorim 87:835–839

Dawodu F, Ayodele O, Xin J, Zhang S, Yan D (2014) Effective conversion of nonedible oil with high free fatty acid into biodiesel by sulphonated carbon catalyst. Appl. Energy 114:819–826

Dehkordi AM, Ghasemi M (2012) Transesterification of waste cooking oil to biodiesel using Ca and Zr mixed oxides as heterogeneous base catalysts. Fuel Process Technol 97:45–51

Demirbas A (2007) Progress and recent trends in biofuels. Prog Energy Combust Sci 33:1–18

Demirbas A (2008) Relationships derived from physical properties of vegetable oil and biodiesel fuels. Fuel 87:1743–1748

Escobar JC, Lora ES, Venturini OJ, Yanez EE, Castillo EF, Almazan O (2009) Biofuels: environment, technology and food security. Renew Sustain Energy Rev 13:1275–1287

Falade OS, Adekunle AS, Aderogba MA, Atanda SO, Harwood C, Adewusi SR (2008) Physicochemical properties, total phenol and tocopherol of some Acacia seed oils. J Sci Food Agric 88:263–268

Freire LMS, Bicudo TC, Rosenhaim R, Sinfrônio FSM, Botelho JR, Carvalho Filho JR, Santos IMG, Fernandes VJ Jr, Antoniosi Filho NR, Souza AG (2009) Thermal investigation of oil and biodiesel from Jatropha curcas L. J Therm Anal Calorim 96:1029–1033

Ganapathy T, Gakkhar RP, Murugesan K (2011) Influence of injection timing on performance, combustion and emission characteristics of Jatropha biodiesel engine. Appl Energy 88:4376–4386

Graboski MS, McCormick RL (1998) Combustion of fat and vegetable oil derived fuels in diesel engines. Prog Energy Combust Sci 24:125–164

Hartikka T, Kuronen M, Kiiski U (2012) Technical performance of HVO (Hydrotreated Vegetable Oil) in diesel engines. SAE technical paper 2012-01-1585. http://dx.doi.org/10.4271/2012-01-1585

Hossain AK, Davies PA (2010) Plant oils as fuels for compression-ignition engines: a technical review and life-cycle analysis. Renew Energy 35:1–13

Huang J, Lin L, Wang Y, Qin J, Roskilly Anthony P, Li L (2011) Experimental study of the performance and emission characteristics of diesel engine using direct and indirect injection systems and different fuels. Fuel Process Technol 92:1380–1386

Ilgen O (2012) Reaction kinetics of dolomite catalyzed transesterification of canola oil and methanol. Fuel Process Technol 95:62–66

International Energy Outlook 2011, Office of communications, EI-40. Washington, DC 20585: U. S. Energy Information Administration, Forrestal Building. DOE/EIA-0484, 2011

Jaichandar S, Senthil Kumar P, Annamalai K (2012) Combined effect to injection timing and combustion chamber geometry on the performance of a bio diesel fueled diesel engine. Energy 47:388–394

Jang MG, Kim DK, Park SC, Lee JS, Kim SW (2012) Biodiesel production from crude canola oil by two-step enzymatic processes. Renew Energy 42:99–104

Joshi G, Pandey JK, Rana S, Rawat DS (2017) Challenges and opportunities for the application of biofuel. Renew Sustain Energy Rev 79:850–866

Kannan GR, Anand R (2011) Experimental evaluation of DI diesel engine operating with diestrol at varying injection pressure and injection timing. Fuel Process Technol 92:2252–2263

Knothe G (2010) Biodiesel and renewable diesel: a comparison. Prog Energy Combust Sci 36:364–3732. https://doi.org/10.1016/j.pecs.2009.11.004

Koyama A, Iki H, Iguchi Y, Tsurutani K (2007) Vegetable oil hydrogenating process for automotive fuel. SAE technical paper 2007-01-2030. http://dx.doi.org/10.4271/2007-01-2030

Kumar Vijay M, Babu Veeresh A, Kumar Ravi P (2017) The impacts on combustion, performance and emissions of biodiesel by using additives in direct injection diesel engine. Alexandria Eng J

Kuti OA, Zhu J, Nishida K, Wang X, Huang Z (2013) Characterization of spray and combustion processes of biodiesel fuel injected by diesel engine common rail system. Fuel 104:838–846

Lapuerta M, Armas O, Rodriguez-Fernandez J (2008) Effect of biodiesel fuels on diesel engine emissions. Prog Energy Combust Sci 34:198–223

Lee HV, Yunus R, Juan JC, Taufiq-Yap YH (2011) Process optimization design for Jatropha-based biodiesel production using response surface methodology. Fuel Process Technol 92:2420–2428

Lotero E, Liu Y, Lopez DE, Suwannakarn K, Bruce DA, Goddwin JG (2005) Synthesis of biodiesel via acid catalysis. Ind Eng Chem Res 44:5353–5363

Marchetti JM, Miguel VU, Errazu AF (2007) Possible methods for biodiesel production. Renew Sustain Energy Rev 11:1300–1311

Masjuki H, Abul K (2013) An overview of biofuel as a renewable energy source: development and challenges. Proc Eng 56:39–53. https://doi.org/10.1016/j.proeng.2013.03.087

Mendonça DR, Andrade HMC, Guimaraes PRB, Vianna RF, Meneghetti SMP, Pontes LAM (2011) Application of full factorial design and Doehlert matrix for the optimization of beef tallow methanolysis via homogeneous catalysis. Fuel Process Technol 3:342–348

Murugesan A, Umarami C, Subramanian R, Nedunchezhian N (2009) Bio-diesel as an alternative fuel for diesel engines—a review. Renew Sustain Energy Rev 13:653–662

Nigam PN, Singh A (2011) Production of liquid biofuels from renewable resources. Prog Energy Combust Sci 37 52–68

No S-Y (2014) Application of hydrotreated vegetable oil from triglyceride based biomass to CI engine—a review. Fuel 115:88–96

Noureddini H, Zhu D (1997) Kinetics of transesterification of soybean oil. J Am Oil Chem Soc 74:1457–1463

Noureddini H, Gao X, Philkana RS (2005) Immobilized Pseudomonas cepacia lipase for biodiesel fuel production from soybean oil. Biores Technol 96:769–777

Nwafor OMI, Rice G, Ogbonna AI (2000) Effect of advanced injection timing on the performance of rapeseed oil in diesel engines. Renew Energy 21:433–444

Pandian M, Sivapirakasam SP, Udayakumar M (2011) Investigationontheeffectof injection system parameters on performance and emission characteristics of a twin cylinder compression ignition direct injection engine fueled with pongamia biodiesel–diesel blend using response surface methodology. Appl Energy 88:2663–2676

Puhan S, Jegan R, Balasubbramanian K, Nagarajan G (2009) Effect of injection pressure on performance, emission and combustion characteristics of high linolenic linseed oil methyl ester in a DI diesel engine. Renew Energy 34:1227–1233

Purushothaman K, Nagarajan G (2009) Effect of injection pressure on heat release rate and emissions in CI engine using orange skin powder diesel solution. Energy Convers Manage 50:962–969

Qi D, Leick M, Liu Y, Chia-fon FL (2011) Effect of EGR and injection timing on combustion and emission characteristics of split injection strategy DI-diesel engine fueled with biodiesel. Fuel 90:1884–1891

Radu R, Petru C, Edward R, Gheorghe M (2009) Fueling an D.I. agricultural diesel engine with waste oil biodiesel: effects over injection, combustion and engine characteristics. Energy Convers Manage 50:2158–2166

Raheman H, Ghadge SV (2008) Performance of diesel engine with biodiesel at varying compression ratio and ignition timing. Fuel 87:2659–2666

Rathore V, Madras G (2007) Synthesis of biodiesel from edible and non-edible oils in supercritical alcohols and enzymatic synthesis in supercritical carbon dioxide. Fuel 86:2650–2659

Salamatinia B, Abdullah AZ, Bhatia S (2012) Quality evaluation of biodiesel produced through ultrasound-assisted heterogeneous catalytic system. Fuel Process Technol 97:1–8

Saravanan AP, Mathimani T, Deviram G, Rajendran K, Pugazhendhi A (2018) Biofuel policy in India: a review of policy barriers in sustainable marketing of biofuel. J Clean Prod 193: 734–747

Shivakumar, Srinivasa Pai P, Shrinivasa Rao BR (2011) Artificial neural network based prediction of performance and emission characteristics of a variable compression ratio CI engine using WCO as a biodiesel at different injection timings. Appl Energy 88:2344–2354

Silva GF, Camargo FL, Ferreira ALO (2011) Application of response surface methodology for optimization of biodiesel production by transesterification of soybean oil with ethanol. Fuel Process Technol 92:407–413

Sugiyama K, Goto I, Kitano K, Mogi K (2009) Effects of hydrotreated vegetable oil (HVO) as renewable diesel fuel on combustion and exhaust emissions in diesel engine. SAE Int J Fuels Lubr 5(1):205–217. https://doi.org/10.4271/2011-01-1954

Van Kasteren JMN, Nisworo AP (2007) A process model to estimate the cost of industrial scale biodiesel production from waste cooking oil by supercritical transesterification. Resour Conserv Recycl 50:442–458

Zhang Y, Yang Y, Boehman AL (2009) Premixed ignition behavior of C9 fatty acid esters: a motored engine study. Combust Flame 156:1202–1213

Zhou W, Boocock DGB (2006) Phase behavior of the base-catalyzed transesterification of soybean oil. J Am Oil Chem Soc 82:1047–1052

Chapter 6
Sustainability Assessment of Biodiesel Production in India from Different Edible Oil Crops Using Emergy Analysis

Shyamal Das, Rahul Dev Misra and Biplab Das

Abstract The demand for energy has increased gradually over the past few years. As a result, the reserve of petroleum-based fuels is depleting continuously which led to thinking about an alternative energy source. This resulted in the increased popularity of biofuel, which is an alternative and renewable source of energy. This study aims to investigate whether biodiesel production from different edible oil crops is sustainable enough to replace conventional fuels. In this work, emergy analysis has been used to study the sustainability of edible oil crops-based biodiesel production. Wheat germ, groundnut and cottonseed oils produced from edible oil crops have been considered in the present study. Based on the data availability, the study area considered in the present work is the state of Maharashtra, India. Three different processes of biodiesel production have been studied in this work, namely alkali-catalysed, acid-catalysed and lipase-catalysed transesterification processes. The emergy performance indicators such as emergy sustainability index (ESI), emergy investment ratio (EIR), emergy renewability (%R), emergy yield ratio (EYR), environmental loading ratio (ELR) and environmental impact ratio (EVR) are then evaluated for each of the considered biodiesels derived from edible oil crops. The sustainability of the considered biodiesels is then evaluated based on a comparative assessment considering these emergy performance indicators. Based on the emergy analysis, it is found that irrespective of the transesterification processes considered, the wheat germ oil is found to be the most preferable edible oil for the production of biodiesel as it has lowest transformity value. On the contrary, cottonseed oil is the least preferred edible oil as it has the highest transformity value. Similarly, based on the values of emergy indices, it is observed that irrespective of the transesterification processes considered, the biodiesel derived from cottonseed oil crop is found to be the most sustainable followed by biodiesel derived from groundnut oil crop and the biodiesel derived from wheat germ oil crop is found to be least sustainable.

S. Das · R. D. Misra · B. Das (✉)
Department of Mechanical Engineering, National Institute of Technology Silchar, Silchar 788010, Assam, India
e-mail: biplab.2kmech@gmail.com

A. K. Agarwal et al. (eds.), *Methanol and the Alternate Fuel Economy*, Energy, Environment, and Sustainability,
https://doi.org/10.1007/978-981-13-3287-6_6

Keywords Biodiesel · Emergy analysis · Emergy performance indicators
Sustainability · Edible oil

6.1 Introduction

Energy is essential for all living beings like plants, animals and human being and man-made systems like vehicles, mechanical equipment. In past years, India has gone through a number of challenges, and security of energy has become a more serious issue. Continuous increase in demand of energy resulted in increased consumption of conventional fuel, mainly fossil fuel, which is non-renewable energy resource and emits greenhouse gases when burned. Thus, the need for alternative source of energy has become very important. In this regard, biomass-derived fuels have gained popularity as an alternative energy source because of their renewability in nature. Fuels obtained from such renewable resources can decrease the need for fossil fuel in the upcoming years. Biodiesel being produced from crops is renewable in nature and has less effect on environment; as a result, it becomes a rapidly growing industry in the recent years (Ong et al. 2012). However, it resulted in several side effects as well, such as increase in food price due to the conversion of a large proportion of food grains into biodiesel, ecological disturbances and the debate of food versus fuel (Gnansounou 2011; Diaz-Chavez 2011).

6.1.1 *Present Worldwide Energy Scenario*

Presently, the increase in energy demand worldwide is directly or indirectly a major challenge for the growth, evolution and survival of the all living beings in the world. It is an important factor in the development of a country socio-economically as well as for the welfare of human beings. Energy has become an important commodity as the supply of which controls the economic development and livelihood of a country. Energy utilization rate of the world is 2.4 kW per person (Smith 2012). The world's demand of energy is gradually increasing, and it is estimated that the energy consumption rate of the world will reach to double of its current value in the upcoming 20 years Indian Petroleum and Natural Gas Statistics (2012). Almost, 80% of the total energy demand of the world is still supplied by fossil fuels.

Increase in motorization and industrialization has led to an increase in consumption of petroleum-based fuels. As there is a limited resource of petroleum-based fuels, the consumption of it at present rate if continues then the time is not far when petroleum-based fuel will become scared. This will lead to an increase in oil price resulting in inflation and thus becoming an obstruction in economic growth.

6.1.2 India's Energy Scenario

For the economic growth of India, the security of energy is of utmost importance. The gap between the energy supply and the energy use can be understood by a proper energy balance. As per statistics, India comes fourth in terms of consumption of oil and petroleum products. It also comes fourth in terms of import of oil and petroleum products. India is highly dependent on import of crude oil. According to Economic Times report, India's import of crude oil has been 184.8 million tonnes (MT) during the financial year of 2012–13, which has increased to 213.93 MT during the financial year of 2016–17 and which further increased to 219.15 MTs during the financial year of 2017–18.

As per the Economic Times report 2017–18, almost 80% of total imported crude oils are used as petrol, diesel and cooking oils. As per the petroleum ministry's working group report, India's demand of crude oil has increased to 205 million tonnes (MT) during the financial year 2017–18. According to Economic Times report, India's production of crude oil has decreased for the sixth straight year in 2017–18 to 35.68 million tonnes (MT) which increased the import of crude oil to 82.8%. Petroleum fuel crisis and increased fossil fuel import are affecting the development and economy of the country. The price increase and environmental effect of fossil fuel led many researchers to go for an alternative and renewable fuel.

To cope up with the current crisis of energy, the main step one can take is to concentrate on the development of technology, which can efficiently utilize the non-conventional source of energy to meet the present fuel requirement of the country.

6.1.3 Alternative and Renewable Fuels

Biofuel. Biofuel is one of the possible energy sources that can solve the problem discussed above. Biofuel is produced from renewable resources by the process of conversion of biomass. Biomass is the living organism, which is either plant or material derived from plant. Biomass can be converted to useful energy by three different processes, namely biochemical conversion, chemical conversion and thermal conversion. The fuels obtained from conversion of biomass are in solid, liquid or gaseous forms. Biodiesel, bio-ethanol and biogas are the different forms of biofuels, and these are called biofuels. The emission from biofuels is considered to be negligible as the carbon dioxide (CO_2) released from burning of biofuels is consumed by plants which are used to produce them. From an environmental perspective, the biofuel is a good alternative for fossil fuels as it results in less emissions of carbon monoxide (CO) and particulates.

Biodiesel. Biodiesel is a form of biofuels produced from domestic renewable resources with minimal emission characteristics. It does not contain petroleum but can be blended with diesel to any amount to form biodiesel blend. Biodiesel can be used as a fuel in diesel engine and oil heating systems without any major change. It has gained a lot of popularity because of its renewable nature, sustainability and lower emissions of greenhouse gases. Biodiesels also have some other characteristics like biodegradability, non-toxicity and free from sulphur as well. In technical term, biodiesel is the mono-alkyl ester of the free fatty acid obtained from animal fats and vegetable oils through esterification process.

Most important difference between biodiesel and diesel is that biodiesel is obtained from plants, seeds and animal fats whereas diesel is obtained from crude oils and refined to different extents. Biodiesel is renewable in nature, whereas diesel is a non-renewable fossil fuel. When biodiesel is used in conventional diesel engine, it reduces the emissions of carbon monoxide, unburned hydrocarbons and particulate emissions as compared to diesel. Biodiesels have higher ozone-forming capability as compared to diesel. Moreover, it reduces the emissions of sulphates and sulphur oxides (SO_X) which are the main cause of acid rain. Nitrogen oxides (NO_X) emissions may increase or decrease depending upon the testing method used and duty cycles of the engine. Soya bean, rapeseed, palm and canola are considered as the first-generation feedstock for the production of biodiesel as they were the first crops to be used for the productions of biodiesel. Similarly, non-food crops like jatropha, karanja are considered as the second-generation feedstock for biodiesel production and algae are considered as a third-generation biodiesel feedstock.

6.2 Research Trend

Different researchers and scientists have developed different methods of biodiesel production from a number of biofuels. It has been reported by many scientists that biodiesel production is more when the process uses a catalyst. Ahn et al. (1995) have studied a two-step reaction process for the production of biodiesel. With the help of this method, different methyl esters such as rapeseed methyl ester, canola methyl ester and sunflower methyl ester have been produced in a reactor using catalysts such as sodium methoxide, potassium hydroxide and sodium hydroxide. Production of biodiesel through two-stage transesterification process using methanol as a catalyst is studied by Cvengro and Povaz (1996). Kaieda et al. (1999) developed a method for the production of methanol and methyl esters of plant oils with the help of enzymatic process in a reaction system free of solvent. Saka and Kusdiana (2001) studied the rapeseed oil transesterification process without using any catalyst. Shimada et al. (2002) have studied the production of biodiesel from biofuel using enzymatic alcoholysis. Ramadhas et al. (2005) have studied the production of biodiesel from rubber seed oil.

In recent years, a number of different methods have been used to determine the sustainability of biodiesel such as multi-criteria decision-making (Nzila et al. 2012; Scott et al. 2012; Mohamadabadi et al. 2009), life cycle assessment (Zhou et al. 2007; Hou et al. 2011; Kim and Dale 2005; Panichelli et al. 2009), ecological footprint (Stoeglehner and Narodoslawsky 2009) and emergy analysis (Ju and Chen

2011; Cavalett and Ortega 2010; Takahasi and Ortega 2010). Out of the methods, emergy analysis is found to be the most suitable method to determine the sustainability of biodiesel as it is a clear and direct method and the parameter emergy sustainability index (ESI) in emergy analysis checks the sustainability of biodiesel produced from different crops. Many researchers have used emergy analysis (EmA) method in order to analyse the production of biodiesel from different biomasses. Nilsson (Nilsson 1997) has studied biomass use as a fuel for district heating plant, and Brown and Ulgiati (2004a) studied the biodiesel production from corn. Similarly, Pereira and Ortega (2010), and Dong (2008) studied the production of biodiesel from sugarcane and wheat. Alonso-Pippo (2004) studied the biodiesel production from sugarcane waste; Sheng et al. Liu et al. (2007) and Cavalett and Ortega (2010) have studied the biofuel production from vegetable oil. Goh and Lee (2010) through their study showed that palm oil generates lower ESI value and commented that Malaysia's 8% of total agricultural area is enough to meet the fuel requirement for the country and palm oil could be a good alternative for conventional fuel. In order to account for the cost benefit of waste management, Mu et al. (2011) incorporated several emergy performance indicators such as ESI, EIR and EYR. These indicators were studied in the production of polythene, and the result so obtained showed that proper waste management leads to a decrease in environmental stress and an increase in sustainability. Nimmanterdwong et al. (2015) in a case study done in Thailand reported that biodiesel produced from oil palm through industrial scale is the most sustainable in nature. A number of investigations have done in recent times to check sustainability of the production of biodiesel. Spinelli et al. (2013) evaluated biodiesel production from sunflower oil using emergy analysis and pointed out some of the limitations in the agricultural phase when it comes to the large scale production of biodiesel from sunflower. Cruz and Nascimento (2012) have carried out the environmental assessment of oil production from microalgae and reported that microalgae are more non-renewable than renewable source for biodiesel feedstock as it has a large raw material consumption, high energy consumption in harvesting and has high capital and operating costs. Biodiesel production from non-edible oil like jatropha has also gained a lot of popularity beside biodiesel production from edible oils. Arbain and Salimon (2011) studied that jatropha is used as a source for biodiesel production in the USA and Europe as it has higher stability due to oxidation than that of soya bean and sunflower oil. Liu et al. (2012) have applied the emergy methods to evaluate the sustainability of petroleum fuel production and rice ethanol production in Japan by analysing the major emergy indices to assess the production processes. They reported that petroleum fuel production presents higher EIR, ELR, EVR and lower EYR, ESI, Tr than rice ethanol production, indicating rice-to-ethanol production makes sense in the reduction of greenhouse gases (GHG). A study in Japan by Hongfang Lu et al. (2012) showed that although ethanol from high-yield rice has a net energy yield, it is currently neither ecologically nor economically feasible to use it to replace gasoline as a transport fuel in Japan. Many biomass sources are available for the production of bio-ethanol such as sugarcane, corn, potato, rice, wheat, as per the study done by both Matsumura et al. (2005) and Hattori and

Morita (2010); out of these, rice is most suitable biomass for the production of biofuel considering both share of land and technology used. Liang et al. (2016) have observed that the production of hydrogen from corn stalks by supercritical water gasification is not sustainable in the long-term perspective in the current situation of Huaibei city in China.

6.3 Emergy Analysis

Emergy is used to compare different forms of energy on a common basis. Suppose it takes 5 kJ of coal energy to generate 1 kJ of electricity (Odum 1988). As these, two are two different forms of energy so they cannot be compared directly, for that one need to express them in some common forms of energy. Since solar energy is the primary or original source for both, solar energy is used as the common energy to compare them in emergy analysis.

Moving across the energy hierarchy, it is seen that the quantity of energy decreases but the quality improves or in other way it can be said that higher amount of lower grade energy is needed to have a relatively lower amount of high-grade energy. Emergy is the method through which these different forms of energy can be expressed on a common unit and can be compared on a quality basis. So, emergy can simply be defined as the energy which is needed to make a product or service. Most commonly used definition is given as:

"Emergy is the form of energy that is needed in the direct or indirect transformation to obtain a product or service" Brown and Ulgiati (2004a).

As discussed earlier, the basic unit of energy is solar energy. So, in emergy analysis, the unit in which the different forms of energy are expressed is solar emijoules (seJ) which implies the amount of solar energy in joule needed to produce a product or service. Though solar emijoules (seJ) is the most common form, there are other emergy units like electrical emijoules and coal emijoules (Brown and Ulgiati 2004a). Sustainability brings forth the question of how the need of human can be fulfilled by generating goods or services having the least effect on the environment. The emergy analysis has been developed by two scientists Odum and Nilsson to evaluate various inputs from human economy and that input which is obtained free from the environment like solar radiation, wind, rain in the production of a good to show the relationship between human need and environmental system (Odum 1988). The emergy method helps in integrating and analysing inputs and outputs to understand the evaluated system. In emergy analysis, both inputs from society and environment which can be called as social and environmental costs are considered unlike energy or exergy analysis. The emergy equivalent of an item or a process is calculated as

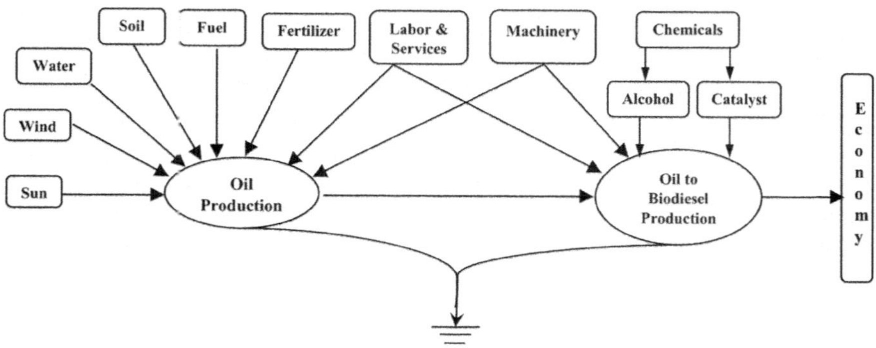

Fig. 6.1 System diagram of emergy flows in the system including all necessary inputs of materials, energy and human services

$$\text{Emergy (seJ)} = \text{Unit} \times \text{Transformity (seJ/unit)} \tag{6.1}$$

The system diagram that shows the flow of emergy of various subsystems is presented in Fig. 6.1. The basis for construction of the diagram is first defining all the input parameters involved and then arranging them in increasing order of transformities from left to right.

6.4 Emergy Performance Indicators

Various emergy-based indicators such as transformity (Tr), emergy investment ratio (EIR), emergy yield ratio (EYR), environmental loading ratio (ELR), emergy sustainability index (ESI), emergy renewability (%R) and environmental impact ratio (EVR) are defined in this section.

Transformity (Tr) value helps in determining the quality of energy. It is defined as the energy input in terms of emergy that is needed to make a unit amount of a product or service. In other words, it is the amount of available energy of lower grade energy needed in the direct or indirect transformation process to get a unit amount of the product. If, for example, to produce 1 J of energy 3000 seJ of wood is needed then the transformity of wood will be 3000 seJ/J. Solar energy is the amplest as well as the most scattered form of energy, and its transformity is considered to be one as per definition Brown and Ulgiati (2004).

There are several types of transformities, and they are used based on their suitability. One such type is specific emergy which is defined as the emergy input per unit mass output and expressed as seJ/g. Transformity can also be expressed in terms of per monetary unit like seJ/$ which helps in evaluating a process involving money exchange. Transactions in terms of money are not paid to the environment directly, so it is evaluated with the help of purchase that is made by the money involved in the transaction. Transformity is also used to describe the quality of the energy. Transformity measures the position of energy in the hierarchy of energy. Transformity value, expressed in seJ/J or seJ/g or seJ/$ units, can be used to estimate the total emergy flow for a product using Eq. (6.2). The emergy flow is the sum of all of the major solar emergy joules used by all of the components of the process. The energy, material or a monetary asset used by a component is multiplied with its corresponding transformity to yield the emergy value for that component in the process. It is estimated over a fixed time period and, therefore, has the unit seJ/time (often seJ/year) (Brown and Ulgiati 2004). The amounts of energy, materials and money used in the calculations are the amounts used over 1 year. A higher value of transformity means a higher amount of solar emijoules is used to make the product.

Transformity of a substance is calculated by dividing the total solar emergy used in the production of the substance by the energy content of the substance.

$$Em = \sum_{i=1}^{n} E_i.Tr_i \tag{6.2}$$

where Em is the total emergy flow or empower, E_i is the amounts of energy or material content used in the calculations, and Tr_i is the corresponding transformity.

Inputs involved in the production process are generally classified as natural inputs or purchased inputs. The natural resources are again of renewable and non-renewable nature. For example, rain, solar radiation are renewable natural resources, while soils and mineral are categorized as non-renewable natural resources. Similarly, the imported resources (I) are classified into purchased inputs (F) and labour and services (S) which in turn are divided into renewable ($F_R + S_R$) and non-renewable fraction ($F_N + S_N$). Labor and services consist of the expenses related to people providing service and fuels, chemicals etc. comes under purchased minerals. Emergy yield (Y) is the emergy of the product obtained in the process.

Emergy investment ratio (EIR) is the ratio of emergy of imported inputs to the emergy of natural inputs. Higher emergy investment ratio means the process is more dependent on purchased resources than natural resources.

$$\text{EIR} = \frac{I}{N+R} = \frac{F+S}{N+R} \tag{6.3}$$

Emergy yield ratio (EYR) is defined as the ratio of total emergy inputs to the imported emergy. It measures the capacity of a process of extracting the resources those are available locally as compared to imported one.

$$\text{EYR} = \frac{Y}{I} = \frac{R+N+I}{I} \tag{6.4}$$

Environmental loading ratio (*ELR*) is defined as the ratio of non-renewable part of natural and purchased resources to the renewable part of purchased and natural resources. A higher value of ELR means that the system has a large impact on the environment.

$$\mathrm{ELR} = \frac{N + F_N + S_N}{R + F_R + S_R} \tag{6.5}$$

Emergy sustainability index (ESI) is defined as the ratio of EYR and ELR. It shows the system's ability to supply emergy to the economy with lowest environmental impact. A system having a higher value of *ESI* is more sustainable.

$$\mathrm{ESI} = \frac{\mathrm{EYR}}{\mathrm{ELR}} \tag{6.6}$$

Emergy renewability (%R) is defined as the renewable fraction of emergy in the total emergy flow. It is an important indicator of process sustainability.

$$\%\mathrm{R} = \frac{R}{Y} \tag{6.7}$$

Environmental impact ratio (*EVR*) is the ratio of external emergy input (EEE) which consists of goods and services to the locally available renewable and non-renewable resources. It measures the impact on the environment due to GHG emission.

$$\mathrm{EVR} = \frac{\mathrm{EEE}}{R + N} \tag{6.8}$$

6.5 Results and Discussion

In this work, emergy analysis has been used to study the sustainability of edible crops-based biodiesel production. Wheat germ, cottonseed and groundnut oils produced from edible oil crops have considered in the present study. Based on the data availability, the study area considered in the present work is the state of Maharashtra, India, and the calculations are performed per hectare and per year basis. Three different processes of biodiesel production are studied—alkali-catalysed transesterification, acid-catalysed transesterification and lipase-catalysed transesterification. The emergy of each item is evaluated by multiplying the available energy of each item with its transformity. Available energy of each item is evaluated based on the collected data from various sources, estimation of certain inputs from the known information and available statistical data of various crops with regard to their production. The details of the results obtained from the emergy analysis of the considered biodiesels with relevant discussions are presented in the following subsections.

6.5.1 Emergy Analysis of Different Crop-Based Bio-oil Productions

The emergy analysis of the considered edible bio-oils, i.e. wheat germ, cottonseed and groundnut oils, has been carried out, and the obtained results are presented in this section. The available energy of the input parameters for each of the considered oils is evaluated using the methodology discussed above. For example, in case of cottonseed oil, available data from the literature shows that an area of 11.69 million hectares can produce cottonseed oil of amount 1.24 million tonnes annually resulting in net oil yield of 157.29 kg/ha/year. Similarly, all the relevant input parameters for each of the considered bio-oils are evaluated. For brevity, the calculations for the available energy of input parameters of cottonseed oil production are shown in Appendix. The results obtained in terms of total emergy and transformity for wheat germ, cottonseed and groundnut oils are presented in Table 6.1.

The various inputs in biodiesel production process consist of natural resources and imported resources. Natural resources consist of renewable and non-renewable fraction, while imported resources are again divided into purchased inputs and labour and services. A fraction of labour and services is considered as renewable. Major inputs from renewable and non-renewable natural resources are solar energy, wind, rain, evapotranspiration and topsoil loss, which consume a considerable fraction of emergy. It can be seen from the tabular results that labour and services consume the maximum portion of emergy for all the different crop-based bio-oil productions. Thus, it implies that bio-oil production process is mainly dependent on economic inputs.

Critical observation of the results clearly demonstrates that groundnut oil would be the better emergy-saving option based on its transformity value (3.14×10^{10}). This is followed by wheat germ and cottonseed oils having transformity values of 7.56×10^{10} and 1.27×10^{11}, respectively. Thus, groundnut oil is the most preferred edible oil for the production of biodiesel because of its lowest transformity value. On the contrary, cottonseed oil is the least preferred edible oil as it has the highest transformity value.

6.5.2 Emergy Analysis of Different Crop-Based Biodiesel Productions

The emergy analysis of the biodiesels produced from the considered edible bio-oils, i.e. wheat germ, cottonseed and groundnut oils, has been carried out, and the obtained results in terms of total emergy and transformity for wheat germ, cottonseed and groundnut oils are presented in Table 6.2.

Critical observation of the results clearly demonstrates that with alkali-catalysed transesterification process, groundnut biodiesel would be the better emergy-saving option based on its transformity value (4.56×10^{10}). This is followed by wheat

Table 6.1 Emergy analysis of wheat germ-, cottonseed- and groundnut-based bio-oil productions

S. No.	Item	Wheat germ oil					Cottonseed oil				
		Quantity	Unit	Transformity (seJ/unit)	Emergy (seJ/ha/year)	% of total emergy used (%)	Quantity	Unit	Transformity (seJ/unit)	Emergy (seJ/ha/year)	% of total emergy used (%)
	Local renewable resources (R)										
1	Solar radiation	6.96E+13	J/ha/year	1.00E+00	6.96E+13	0.65	6.96E+13	J/ha/year	1.00E+00	6.96E+13	0.81
2	Wind energy	6.34E+10	J/ha/year	2.45E+03	1.55E+14	1.45	6.34E+10	J/ha/year	2.45E+03	1.55E+14	1.45
3	Rain (chemical potential)	5.69E+10	J/ha/year	3.06E+04	1.74E+15	16.29	5.69E+10	J/ha/year	3.06E+04	1.74E+15	16.29
4	Evapotranspiration	9.29E+10	J/ha/year	3.06E+04	2.84E+15	25.74	9.29E+10	J/ha/year	3.06E+04	2.84E+15	21.09
	Local non-renewable resources (N)										
5	Net topsoil loss (organic matter)	2.39E+09	J/ha/year	2.25E+05	5.39E+14	5.05	2.39E+09	J/ha/year	2.25E+05	5.39E+14	5.05
	Purchased resources (F)										
6	Seeds	1.25E+05	g/ha/year	3.98E+10	4.98E+15	0.00	1.20E+04	gfaa/year	5.61E+10	6.74E+14	0.00
7	Fertilizers: nitrogen (N)	5.57E+04	g/ha/year	6.38E+09	3.55E+14	3.22	7.58E+04	gfaa/year	6.38E+09	4.84E+14	3.59
8	Phosphorus (P_2O_5)	1.59E+04	g/ha/year	6.55E+09	1.04E+14	0.94	1.82E+04	gha/year	6.55E+09	1.19E+14	0.88
9	Potassium (K_2O)	4.30E+03	g/ha/year	1.85E+09	7.96E+12	0.07	3.60E+03	gha/year	1.85E+09	6.66E+12	0.05
10	Diesel	2.10E+04	Rs/ha/year	1.10E+05	2.31E+09	0.00	3.39E+04	Rs/ha/year	1.10E+05	3.73E+09	0.00
	Labour and services (S)										
11	Herbicides, pesticides	7.15E+03	g/ha/year	2.49E+10	1.78E+14	1.61	1.45E+03	gha/year	2.49E+10	3.62E+13	0.27
12	Labour	1.20E+04	Rs/ha/year	2.69E+11	3.22E+15	29.18	2.43E+04	Rs/ha/year	2.69E+11	6.55E+15	48.57
13	Machinery	6.80E+03	Rs/ha/year	2.69E+11	1.83E+15	16.56	3.49E+03	Rs/ha/year	2.69E+11	9.40E+14	6.97

(continued)

Table 6.1 (continued)

S. No.	Item	Wheat germ oil					Cottonseed oil				
		Quantity	Unit	Transformity (seJ/unit)	Emergy (seJ/ha/year)	% of total emergy used (%)	Quantity	Unit	Transformity (seJ/unit)	Emergy (seJ/ha/year)	% of total emergy used (%)
	Product:										
	Crude oil	1.46E+05	g/ha/year				1.06E+05	gha/year			
	Total emergy (w/o L&S)				5.81E+15	52.65				5.96E+15	44.19
	Total emergy (with L&S)				1.10E+16	100.00				1.35E+16	100.00
	Transformity (w/o L&S)			3.98E+10					5.61E+10		
	Transformity (with L&S)			7.56E+10					1.27E+11		

S. No.	Item	Groundnut oil					Reference for input quantity	Reference for transformity
		Quantity	Unit	Transformity (seJ/unit)	Emergy (seJ/ha/year)	% of total emergy used (%)		
	Local renewable resources (R)							
1	Solar radiation	6.96E+13	J/ha/year	1.00E+00	6.96E+13	0.50	(1)	Definition
2	Wind energy	6.34E+10	J/ha/year	2.45E+03	1.55E+14	1.45	(1)	(9)
3	Rain (Chemical potential)	5.69E+10	J/ha/year	3.06E+04	1.74E+15	16.29	(1)	(10)

(continued)

Table 6.1 (continued)

S. No.	Item	Groundnut oil					Reference for input quantity	Reference for transformity
		Quantity	Unit	Transformity (seJ/unit)	Emergy (seJ/ha/year)	% of total emergy used (%)		
4	Evapotranspiration	8.48E+10	J/ha/year	3.06E+04	2.60E+15	20.55	(2), (3)	(11)
	Local non-renewable resources (N)							
5	Net topsoil loss (organic matter)	2.39E+09	J/ha/year	2.25E+05	5.39E+14	5.05	(4)	(11)
	Purchased resources (F)							
6	Seeds	1.00E+05	g/ha/year	1.36E+10	1.36E+15	0.00	(5)	This study
7	Fertilizers: nitrogen (N)	2.19E+04	g/ha/year	6.38E+09	1.40E+14	1.11	(6)	(10)
8	Phosphorus (P_2O_5)	3.60E+04	g/ha/year	6.55E+09	2.36E+14	1.87	(6)	(10)
9	Potassium (K_2O)	9.20E+03	g/ha/year	1.85E+09	1.70E+13	0.13	(6)	(10)
10	Diesel	2.71E+04	Rs/ha/year	1.10E+05	2.98E+09	0.00	(7), (8)	(11)
	Labour and services (S)							
11	Herbicides, pesticides	2.91E+03	g/ha/year	2.49E+10	7.26E+13	0.57	(5)	(11)
12	Labour	2.28E+04	Rs/ha/year	2.69E+11	6.12E+15	48.50	(8)	(12)
13	Machinery	3.49E+03	Rs/ha/year	2.69E+11	9.38E+14	7.43	(8)	(12)
	Product:							
	crude oil	4.02E+05	g/ha/year				(5)	
	Total emergy (w/o L&S)				5.49E+15	43.49		

(continued)

Table 6.1 (continued)

S. No.	Item	Groundnut oil					Reference for input quantity	Reference for transformity
		Quantity	Unit	Transformity (seJ/unit)	Emergy (seJ/ha/year)	% of total emergy used (%)		
	Total emergy (with L&S)				1.26E+16	100.00		
	Transformity (w/o L&S)			1.36E+10				
	Transformity (with L&S)			3.14E+10				

(1) Atmospheric Science Data Center. NASA Surface meteorology and Solar Energy. Available at: https://eosweb.larc.nasa.gov/cgi-bin/sse/grid.cgi?email=skip@larc.nasa.gov

(2) Meteorological datasets- India Water Portal. Available at: http://www.indiawaterportal.org/met_data

(3) FAO (Food and Agricultural Organisation). Single crop coefficients. Available at: http://www.fao.org/docrep/X0490E/x0490e0b.htm#crop coefficients

(4) Agricultural Science, Farm Technology, Cultivation method. Available at: http://www.thehindu.com/sci-tech/agriculture/India-losing-5334-million-tonnes-of-soil-annually due-to-erosion-Govt

(5) Directorate of Economics and Satistics, Department of Agriculture and Co-operation, Ministry of Agriculture, Government of India

(6) FAO (Food and Agricultural Organisation). Fertilizer use by crop in India. Available at: http://www.fao.org/docrep/009/a0257e/A0257E05.htm

(7) Direct and indirect use of fossils fuels in farming: cost of fuel price rise for Indian agriculture, 2014

(8) Commission for agricultural cost and prices. Ministry of agriculture and farmer's welfare, Government of India

(9) Odum (2000)

(10) Odum (1996)

(11) Brown Ulgiati (2004b)

(12) The National Environmental Accounting Database (NEAD) (2008)

Table 6.2 Emergy analysis of wheat, cottonseed and groundnut biodiesel production (a. base-catalysed, b. acid-catalysed, c. lipase-catalysed)

S. no.	Item	Wheat germ oil					Cottonseed oil				
		Quantity	Unit	Transformity (seJ/unit)	Emergy (seJ/ha/year)	% of total emergy used (%)	Quantity	Unit	Transformity (seJ/unit)	Emergy (seJ/ha/year)	% of total emergy used (%)
1	Oil from agricultural phase (w/o L&S)	1.46E+05	g/ha/year	3.98E+10	5.81E+15		1.06E+05	gha/year	5.61E+10	5.96E+15	
2	Oil from agricultural phase (with L&S)	1.46E+05	g/ha/year	7.56E+10	1.10E+16		1.06E+05	gha/year	1.27E+11	1.35E+16	
3	Water	6.49E+08	g/ha/year	1.25E+06	8.11E+14	5.67	8.86E+08	gha/year	1.25E+06	1.11E+15	6.16
	Purchased resources (F)										
4	Electricity	4.32E+08	Rs/ha/year	3.36E+05	1.45E+14	1.01	4.32E+08	Rs/ha/year	3.36E+05	1.45E+14	0.81
5a	Methanol	5.15E+04	g/ha/year	7.23E+09	3.72E+14	2.60	5.15E+04	gha/year	7.23E+09	3.72E+14	2.07
5b	Methanol	3.06E+06	g/ha/year	7.23E+09	2.21E+16	43.24	2.30E+05	gha/year	7.23E+09	1.66E+15	9.44
5c	Ethanol	2.35E+05	g/ha/year	7.23E+09	1.70E+15	7.68	1.77E+04	gha/year	7.23E+09	1.28E+14	0.83
6a	Sodium hydroxide (NaOH)	2.50E+03	g/ha/year	6.38E+09	1.60E+13	0.11	2.50E+03	gha/year	6.38E+09	1.6E+13	0.11
6b	Sulphuric acid (H_2SO_4)	1.46E+06	g/ha/year	6.38E+09	9.31E+15	18.23	1.06E+05	gha/year	6.38E+09	6.77E+14	3.84
6c	Lypozyme	1.20E+05	g/ha/year	6.38E+09	7.63E+14	3.44	9.01E+03	gha/year	6.38E+09	5.75E+13	0.37
	Labour and services (S)										
7	Labour	1.20E+04	Rs/ha/year	1.03E+11	1.23E+15	8.59	2.43E+04	Rs/ha/year	1.03E+11	2.50E+15	13.90
8	Machinery	6.80E+03	Rs/ha/year	1.03E+11	6.98E+14	4.88	3.49E+03	Rs/ha/year	2.69E+11	9.40E+14	6.97
	Product:										
9a	Biodiesel	1.39E+05	g/ha/year				1.01E+05	gha/year			
9b	Biodiesel	8.76E+05	g/ha/year				6.36E+04	gha/year			
9c	Biodiesel	1.21E+06	g/ha/year				8.80E+04	gha/year			

(continued)

Table 6.2 (continued)

S. no.	Item	Wheat germ oil					Cottonseed oil				
		Quantity	Unit	Transformity (seJ/unit)	Emergy (seJ/ha/year)	% of total emergy used (%)	Quantity	Unit	Transformity (seJ/unit)	Emergy (seJ/ha/year)	% of total emergy used (%)
10a	Total emergy (w/o L&S)				7.16E+15	50.00				7.6E+15	42.26
10b	Total emergy (w/o L&S)				4.55E+16	89.01				9.55E+15	54.13
10c	Total emergy (w/o L&S)				1.65E+16	74.64				7.39E+15	47.75
11a	Total emergy (with L&S)				1.43E+16	100.00				1.8E+16	100.00
11b	Total emergy (with L&S)				5.11E+16	100.00				1.76E+16	100.00
11c	Total emergy (with L&S)				2.22E+16	100.00				1.55E+16	100.00
12a	Transformity (w/o L&S)			5.16E+10					7.54E+10		
12b	Transformity (w/o L&S)			5.19E+10					1.50E+11		
12c	Transformity (w/o L&S)			1.36E+10					8.40E+10		
13a	Transformity (with L&S)			1.03E+11					1.78E+11		
13b	Transformity (with L&S)			5.83E+10					2.77E+11		
13c	Transformity (with L&S)			1.83E+10					1.76E+11		

S. no.	Item	Groundnut oil					Reference for input quantity	Reference for transformity
		Quantity	Unit	Transformity (seJ/unit)	Emergy (seJ/ha/year)	% of total emergy used (%)		
1	Oil from agricultural phase (w/o L&S)	4.02E+05	g/ha/year	1.36E+10	5.49E+15			
2	Oil from agricultural phase (with L&S)	4.02E+05	g/ha/year	3.14E+10	1.26E+16			
3	Water	1.26E+09	g/ha/year	1.25E+06	1.58E+15	9.06	(1)	(7)
	Purchased resources (F)							

(continued)

Table 6.2 (continued)

S. no.	Item	Groundnut oil					Reference for input quantity	Reference for transformity
		Quantity	Unit	Transformity (seJ/unit)	Emergy (seJ/ha/year)	% of total emergy used (%)		
4	Electricity	4.32E+08	Rs/ha/year	3.36E+05	1.45E+14	0.83	(2)	(8)
5a	Methanol	5.15E+04	g/ha/year	7.23E+09	0.021344	2.68	(3)	(9)
5b	Methanol	8.27E+05	g/ha/year	7.23E+09	5.98E+15	25.06	(4)	(9)
5c	Ethanol	6.37E+04	g/ha/year	7.23E+09	4.60E+14	2.88	(5)	(9)
6a	Sodium hydroxide (NaOH)	2.50E+03	g/ha/year	6.38E+09	1.6E+13	0.11	(3)	(8)
6b	Sulphuric acid (H_2SO_4)	4.02E+05	g/ha/year	6.38E+09	2.57E+15	10.76	(4)	(8)
6c	Lypozyme	3.23E+04	g/ha/year	6.38E+09	2.06E+14	1.29	(5)	(8)
	Labour and services (S)							
7	Labour	2.28E+04	Rs/ha/year	1.03E+11	2.34E+15	13.41	(6)	(10)
8	Machinery	3.49E+03	Rs/ha/year	2.69E+11	9.38E+14	7.43	(6)	(10)
	Product:							
9a	Biodiesel	3.82E+05	g/ha/year				(3)	
9b	Biodiesel	2.41E+05	g/ha/year				(4)	
9c	Biodiesel	3.34E+05	g/ha/year				(5)	
10a	Total emergy (w/o L&S)				7.61E+15	43.61		
10b	Total emergy (w/o L&S)				1.58E+16	66.08		
10c	Total emergy (w/o L&S)				7.88E+15	49.36		
11a	Total emergy (with L&S)				1.74E+16	100.00		
11b	Total emergy (with L&S)				2.39E+16	100.00		
11c	Total emergy (with L&S)				1.6E+16	100.00		
12a	Transformity (w/o L&S)			1.99E+10				
12b	Transformity (w/o L&S)			6.53E+10				

(continued)

Table 6.2 (continued)

S. no.	Item	Groundnut oil					Reference for input quantity	Reference for transformity
		Quantity	Unit	Transformity (seJ/unit)	Emergy (seJ/ha/year)	% of total emergy used (%)		
12c	Transformity (w/o L&S)			2.36E+10				
13a	Transformity (with L&S)			4.56E+10				
13b	Transformity (with L&S)			9.88E+10				
13c	Transformity (with L&S)			4.78E+10				

(1) Water Stats- Water Footprint Statistics- Water footprint network. Available at: http://waterfootprint.org/en/resources/water-footprint-statistics/
(2) Directorate of Economics and Satistics, Department of Agriculture and Co-operation, Ministry of Agriculture, Government of India
(3) Jena and Misra (2014)
(4) Miao and Wu (2006)
(5) Selmi and Thomas (1998)
(6) Commission for agricultural cost and prices. Ministry of agriculture and farmer's welfare, Government of India
(7) Brown and Arding (1991)
(8) Odum (1996)
(9) Cao and Feng (Cao and Feng 2008)
(10) The National Environmental Accounting Database (NEAD) (2008)

germ and cottonseed biodiesels having transformity values of 1.03×10^{11} and 1.78×10^{11}, respectively. Again with acid-catalysed transesterification process, wheat germ biodiesel would be the better emergy-saving option based on its transformity value (5.83×10^{10}). This is followed by groundnut and cottonseed biodiesels having transformity values of 9.88×10^{10} and 2.77×10^{11}, respectively. Further with lipase-catalysed transesterification process, wheat germ biodiesel would be the better emergy-saving option based on its transformity value (1.83×10^{10}). This is followed by groundnut and cottonseed biodiesels having transformity values of 4.78×10^{10} and 1.76×10^{11}, respectively.

Thus, groundnut biodiesel is the most preferred biodiesel, produced through alkali-catalysed transesterification process, because of its lowest transformity value. On the contrary, cottonseed biodiesel is the least preferred biodiesel as it has the highest transformity value. Again, wheat germ biodiesel is the most preferred biodiesel, produced through acid-catalysed transesterification process, because of its lowest transformity value. On the other hand, cottonseed biodiesel is the least preferred biodiesel as it has the highest transformity value. Further, wheat germ biodiesel is the most preferred biodiesel, produced through lipase-catalysed transesterification process, because of its lowest transformity value. On the contrary, cottonseed biodiesel is the least preferred biodiesel as it has the highest transformity value.

6.5.3 Comparative Assessment of Different Biodiesels Based on Emergy Performance Indicators

In order to evaluate the sustainability of the considered biodiesels produced through different transesterification processes, their emergy performance is evaluated using different emergy performance indicators such as emergy investment ratio (EIR), emergy yield ratio (EYR), environmental loading ratio (*ELR*), emergy sustainability index (*ESI*), emergy renewability (%R) and environmental impact ratio (*EVR*). For comparative assessment, the values of these emergy performance indicators are plotted against each of the considered biodiesels. Figures 6.2, 6.3 and 6.4 show the results for alkali-, acid- and lipase-catalysed transesterification processes, respectively. In the comparative assessment, the-larger-the-better criteria are followed for ESI, EYR and %R, whereas the-smaller-the-better criteria are followed for ELR, EIR and EVR. This is as per the definition of these performance indicators.

Emergy Performance Indicators for Alkali-Catalysed Edible Oil-based Biodiesel Evaluation of emergy performance indicators of the biodiesels produced with alkali-catalysed transesterification process from the considered edible bio-oils, i.e. wheat germ, cottonseed and groundnut oils, has been carried out, and the obtained results are presented in this section. The results obtained in terms of emergy performance indicators for wheat germ, cottonseed and groundnut oils are presented in Fig. 6.2.

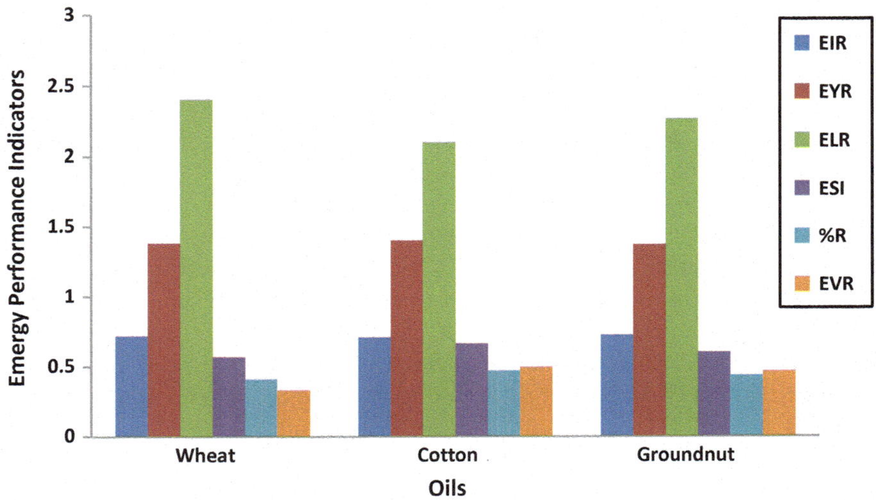

Fig. 6.2 Emergy indices versus edible oils for alkali-catalysed process

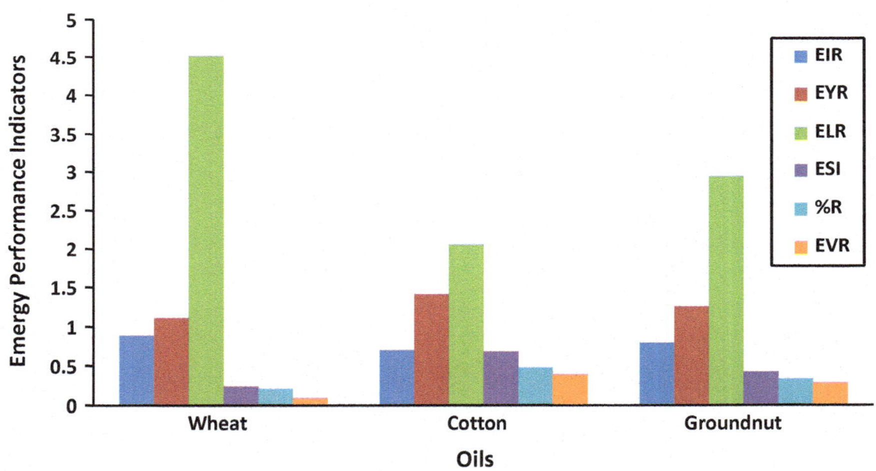

Fig. 6.3 Emergy indices versus edible oils for acid-catalysed process

Critical observation of the results clearly demonstrates that with alkali-catalysed transesterification process, cottonseed biodiesel has the lowest EIR value (0.713), so the share of economic input in total solar emijoules is the lowest for cottonseed biodiesel. This is followed by wheat germ (0.722) and groundnut (0.728) biodiesels. Similarly, based on *%R* value, cottonseed biodiesel (0.475) would be the better as it shows the highest renewability followed by groundnut (0.441) and wheat germ (0.414) biodiesel. Again, based on EYR value, cottonseed biodiesel (1.401) would

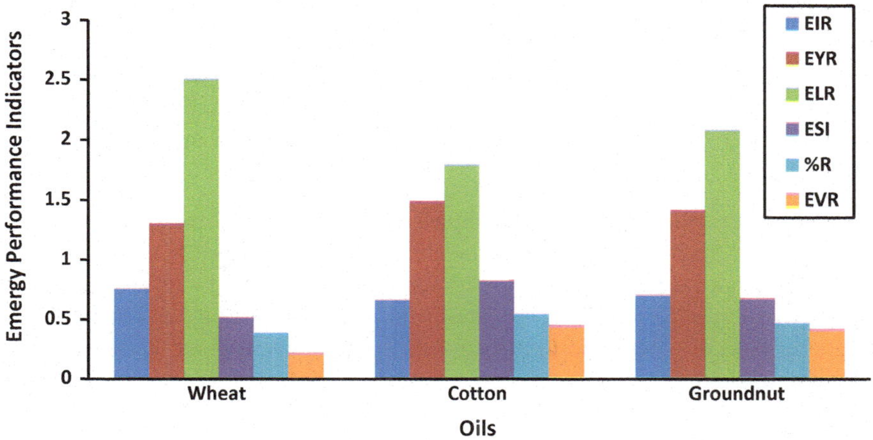

Fig. 6.4 Emergy indices versus edible oils for lipase-catalysed process

be the better, thus requiring the lowest purchased resources input. This is followed by wheat germ (1.383) and groundnut (1.372) biodiesels. Further based on *ELR* value, cottonseed biodiesel (2.103) would be the better as it shows the lowest disturbances to the local environment. This is followed by groundnut (2.267) and wheat germ (2.410) biodiesels. Again based on *EVR* value, wheat germ (0.334) would be better as it has the lowest environmental impact due to GHG emission. This is followed by groundnut (0.471) and cottonseed (0.501) biodiesels. Accordingly, *ESI* value of cottonseed biodiesel (0.666) is found to be the highest and as such it is the most sustainable biodiesel followed by groundnut (0.605) and wheat germ (0.574) biodiesels.

Emergy Performance Indicators for Acid-Catalysed Edible Oil-based Biodiesel Evaluation of emergy performance indicators of the biodiesels produced with acid-catalysed transesterification process from the considered edible bio-oils, i.e. wheat germ, cottonseed and groundnut oils, has been carried out, and the obtained results in terms of emergy performance indicators for wheat germ, cottonseed and groundnut oils are presented in Fig. 6.3.

Critical observation of the results clearly demonstrates that with acid-catalysed transesterification process, cottonseed biodiesel has the lowest EIR value (0.708), so the share of economic input in total solar emijoules is the lowest for cottonseed biodiesel. This is followed by groundnut (0.797) and wheat germ (0.896) biodiesels. Similarly, based on *%R*, cottonseed biodiesels (0.484) would be the better as it shows the highest renewability followed by groundnut (0.339) and wheat germ (0.221) biodiesel. Again based on EYR value, cottonseed biodiesel (1.412) would be the better, thus requiring the lowest purchased resources input. This is followed by groundnut (1.253) and wheat germ (1.115) biodiesels. Further based on *ELR* value, cottonseed biodiesel (2.064) would be the better as it shows the lowest disturbances to the local environment. This is followed by groundnut (2.942) and wheat germ (4.526) biodiesels. Again based on *EVR* value, wheat germ (0.197) would be better as it has the lowest environmental impact due to GHG emission. This is followed by groundnut (0.288) and cottonseed (0.325) biodiesels. Accordingly, *ESI* value of cottonseed biodiesel (0.684) is found to be the highest and as such it is the most sustainable biodiesel followed by groundnut (0.426) and wheat germ (0.246) biodiesels.

Emergy performance indicators for lipase-catalysed edible oil-based biodiesel Evaluation of emergy performance indicators of the biodiesels produced with lipase-catalysed transesterification process from the considered edible bio-oils, i.e. wheat germ, cottonseed and groundnut oils, has been carried out, and the obtained results in terms of emergy performance indicators for wheat germ, cottonseed and groundnut oils are presented in Fig. 6.4.

Critical observation of the results clearly demonstrates that with lipase-catalysed transesterification process, cottonseed biodiesel has the lowest EIR value (0.669), so the share of economic input in total solar emijoules is the lowest for cottonseed biodiesel. This is followed by groundnut (0.705) and wheat germ (0.763) biodiesels. Similarly, based on %R, cottonseed biodiesels (0.554) would be the better as it shows the highest renewability followed by groundnut (0.478) and wheat germ (0.397) biodiesel. Again based on EYR value, cottonseed biodiesel (1.494) would be the better, thus requiring the lowest purchased resources input. This is followed by groundnut (1.416) and wheat germ (1.308) biodiesels. Further based on ELR value, cottonseed biodiesel (1.805) would be the better as it shows the lowest disturbances to the local environment. This is followed by groundnut (2.091) and wheat germ (2.516) biodiesels. Again based on EVR value, wheat germ (0.258) would be better as it has the lowest environmental impact due to GHG emission. This is followed by groundnut (0.419) and cottonseed (0.450) biodiesels. Accordingly, ESI value of cottonseed biodiesel (0.828) is found to be the highest and as such it is the most sustainable biodiesel followed by groundnut (0.677) and wheat germ (0.520) biodiesels.

Overall Discussion Among the emergy performance indicators, the *ESI* indicator represents the sustainability and EIR represents total emergy invested. The biodiesel having the highest sustainability with the lowest emergy investment is naturally the most preferred biodiesel. Similarly, *%R* represents fraction renewability and *EVR* represents environmental impact. Thus, biodiesel having higher renewability and lower environmental impact will be the most preferable. An analysis has been carried out to list down the considered biodiesels in a chronological order of preference from highest to lowest.

Based on the analysis and discussions presented in the preceding sections, the biodiesels derived from various edible oil-based crops can be tabulated as per their selectivity considering ESI, %R, EIR and EVR indicators. These results are presented in Table 6.3 for all the considered transesterification processes (i.e. alkali-, acid- and lipase-catalysed transesterification processes). Finally, the considered biodiesels derived from edible oil-based crops are listed on the basis of their preference and are presented in Table 6.4.

Table 6.3 Selectivity order of biodiesel from edible oil for different transesterification processes

		Selectivity order		
		1	2	3
Alkali	ESI	Cotton	Groundnut	Wheat
	%R	Cotton	Groundnut	Wheat
	EIR	Cotton	Wheat	Groundnut
	EVR	Wheat	Groundnut	Cotton
Acid	ESI	Cotton	Groundnut	Wheat
	%R	Cotton	Groundnut	Wheat
	EIR	Cotton	Groundnut	Wheat
	EVR	Wheat	Groundnut	Cotton
Lipase	ESI	Cotton	Groundnut	Wheat
	%R	Cotton	Groundnut	Wheat
	EIR	Cotton	Groundnut	Wheat
	EVR	Wheat	Groundnut	Cotton

Table 6.4 Preference order of biodiesel from edible oil for different transesterification processes

Preference order	Alkali	Acid	Lipase
1	Cotton	Cotton	Cotton
2	Groundnut	Groundnut	Groundnut
3	Wheat	Wheat	Wheat

These tabular results for alkali-catalysed transesterification process clearly depict that biodiesels derived from cottonseed oil are the most preferable followed by biodiesels derived from groundnut and wheat germ. Similarly, for acid-catalysed transesterification process, biodiesels derived from cottonseed oil are the most preferable followed by biodiesels derived from groundnut and wheat germ. Again, for lipase-catalysed transesterification process, biodiesels derived from cottonseed oil are the most preferable followed by biodiesels derived from groundnut and wheat germ.

6.6 Summary

Various outcomes from the emergy analysis of the considered edible biodiesels can be summarized as follows:

- From transformity point of view:
  - With alkali-catalysed transesterification process, groundnut biodiesel would be the better emergy-saving option based on its transformity value (4.56×10^{10}). This is followed by wheat germ and cottonseed biodiesels having transformity values of 1.03×10^{11} and 1.78×10^{11}, respectively.

  - With acid-catalysed transesterification process, wheat germ biodiesel would be the better emergy-saving option based on its transformity value (5.83×10^{10}). This is followed by groundnut and cottonseed biodiesels having transformity values of 9.88×10^{10} and 2.77×10^{11}, respectively.
  - With lipase-catalysed transesterification process, wheat germ biodiesel would be the better emergy-saving option based on its transformity value (1.83×10^{10}). This is followed by groundnut and cottonseed biodiesels having transformity values of 4.78×10^{10} and 1.76×10^{11}, respectively.

- From emergy investment point of view:

  - With alkali-catalysed transesterification process, the EIR value of cottonseed crop is found to be lower than wheat germ and groundnut by 1.26 and 2.10%, respectively. Hence, the biodiesel derived from cottonseed crop is found to be the most preferable because it requires minimum emergy investment. This is followed by the biodiesels derived from wheat germ. The biodiesel derived from groundnut is the least preferred.
  - With acid-catalysed transesterification process, the EIR value of cottonseed crop is found to be lower than groundnut and wheat germ by 1.25 and 2.65%, respectively. Hence, the biodiesel derived from cottonseed crop is found to be the most preferable, followed by the biodiesels derived from groundnut. The biodiesel derived from wheat germ is the least preferred.
  - With lipase-catalysed transesterification process, the EIR value of cottonseed crop is found to be lower than groundnut and wheat germ by 5.38 and 14.05%, respectively. Hence, the biodiesel derived from cottonseed crop is found to be the most preferable, followed by the biodiesels derived from groundnut. The biodiesel derived from wheat germ is the least preferred.

- From emergy sustainability point of view:

  - With alkali-catalysed transesterification process, the *ESI* value of cottonseed crop is found to be higher than groundnut and wheat germ by 9.16 and 13.81%, respectively. Hence, the biodiesel derived from cottonseed crop is found to be the most preferable because it has maximum sustainability value. This is followed by the biodiesels derived from groundnut. The biodiesel derived from wheat germ is the least preferred.
  - With acid-catalysed transesterification process, the *ESI* value of cottonseed crop is found to be higher than groundnut and wheat germ by 37.71 and 64.03%, respectively. Hence, the biodiesel derived from cottonseed crop is found to be the most preferable, followed by the biodiesels derived from groundnut. The biodiesel derived from wheat germ is the least preferred.
  - With lipase-catalysed transesterification process, the *ESI* value of cottonseed crop is found to be higher than groundnut and wheat germ by 18.23 and 37.19%, respectively. Hence, the biodiesel derived from cottonseed crop is found to be the most preferable, followed by the biodiesels derived from groundnut. The biodiesel derived from wheat germ is the least preferred.

- From emergy renewability point of view:
  - With alkali-catalysed transesterification process, the *%R* value of cottonseed crop is found to be higher than groundnut and wheat germ by 7.16 and 12.84%, respectively. Hence, the biodiesel derived from cottonseed crop is found to be the most preferable because it has the highest renewable fraction of emergy flow. This is followed by the biodiesels derived from groundnut. The biodiesel derived from wheat germ is the least preferred.
  - With acid-catalysed transesterification process, the *%R* value of cottonseed crop is found to be higher than groundnut and wheat germ by 29.95 and 54.33%, respectively. Hence, the biodiesel derived from cottonseed crop is found to be the most preferable, followed by the biodiesels derived from groundnut. The biodiesel derived from wheat germ is the least preferred.
  - With lipase-catalysed transesterification process, the *%R* value of cottonseed crop is found to be higher than groundnut and wheat germ by 13.72 and 28.34%, respectively. Hence, the biodiesel derived from cottonseed crop is found to be the most preferable, followed by the biodiesels derived from groundnut. The biodiesel derived from wheat germ is the least preferred.
- From environmental impact ratio point of view:
  - With alkali-catalysed transesterification process, the *EVR* value of wheat germ crop is found to be lower than groundnut and cottonseed by 41.02 and 50%, respectively. Hence, the biodiesel derived from wheat crop is found to be the most preferable because it has the lowest environmental impact due to GHG emission. This is followed by the biodiesels derived from groundnut. The biodiesel derived from cottonseed is the least preferred.
  - With acid-catalysed transesterification process, the *EVR* value of wheat germ crop is found to be lower than groundnut and cottonseed by 46.19 and 64.97%, respectively. Hence, the biodiesel derived from wheat crop is found to be the most preferable, followed by the biodiesels derived from groundnut. The biodiesel derived from cottonseed is the least preferred.
  - With lipase-catalysed transesterification process, the *EVR* value of wheat germ crop is found to be lower than groundnut and cottonseed by 62.40 and 74.42%, respectively. Hence, the biodiesel derived from wheat crop is found to be the most preferable, followed by the biodiesels derived from groundnut. The biodiesel derived from cottonseed is the least preferred.

6.7 Conclusions

For biodiesels derived from different edible oils, the following conclusions are drawn from the emergy analysis:

- With alkali-catalysed transesterification process, the biodiesel derived from cottonseed crop is found to be the most preferable. This is followed by the biodiesels derived from groundnut, and the biodiesel derived from wheat germ is the least preferred.
- With acid-catalysed transesterification process, the biodiesel derived from cottonseed crop is found to be the most preferable, followed by the biodiesels derived from groundnut, and the biodiesel derived from wheat germ is the least preferred.
- With lipase-catalysed transesterification process, the biodiesel derived from cottonseed crop is found to be the most preferable, followed by the biodiesels derived from groundnut, and the biodiesel derived from wheat germ is the least preferred.
- In overall irrespective of the emergy performance indicators and transesterification processes considered, the biodiesel derived from cottonseed crop is found to be the most preferable, followed by the biodiesels derived from groundnut. The biodiesel derived from wheat germ is the least preferred.

Appendix: Available Energy of Input Parameters for Cottonseed Biodiesel Production: A Sample Calculation

	Area = 1 ha = 10^4 m^2
Sunlight	Solar insolation = 5.2983 kwh/m^2/day
	Energy = 5.2983 × 3600 × 10^3 × 365 × 10^4 = 6.962 × 10^{13} J/ha/year
Rain chemical potential	Area = 1 ha = 10^4 m^2
	Average rainfall = 1.151 m/year
	Density = 1000 kg/m3
	Gibbs free energy = 4940 J/kg
	Energy = average rainfall × area × density × gibbs free energy = 5.6859 × 10^{10}
Wind	Avg. wind speed = 3.28 m/sec
	Air density = 1.23 kg/m^3
	Area = 1 ha = 10^4 m^2
	Drag coefficient = 0.001
	Avg. wind speed = 0.6 × geostrophic wind
	Energy = area × air density × drag coefficient × (geostrophic wind) × time
Net topsoil loss (NTL)	Avg. soil loss = 16.4 tonnes/ha/year
	Organic matter content = 1–6%
	Energy content = 1.46 × 10^{10} J/tonnes
	Energy of NTL = soil loss × organic matter content × energy content = 2.394 × 10^9

References

Ahn E, Koncar M, Mittelbach M, Man R (1995) A low-waste process for the production of biodiesel. Sci Technol 30:2021–2033

Alonso-Pippo W (2004) Emergy evaluation of bio-oil production using sugarcane biomass residues at fast pyrolysis pilot in Brazil In: Proceedings of the IV biennial international workshop "advances in energy studies". Unicamp, Campinas, Brazil

Arbain NH, Salimon J (2011) Synthesis and characterization of ester trimethylolpropane based jatropha curcas oil as biolubricant base stock. J Sci Technol 2(2)

Brown MT, Arding J (1991) Transformity working paper. Center for wetlands, University of Florida, Gainesville, USA

Brown MT, Ulgiati S (2004a) Energy quality, emergy, and transformity: H.T. Odum's contributions to quantifying and understanding systems. Ecol Model 178(2):201–213

Brown MT, Ulgiati S (2004b) Emergy analysis and environmental accounting. Encyclopedia of Energy 2:329–354

Cao K, Feng X (2008) The emergy analysis of loop circuit. Env Monit Assess 147:243–251

Cavalett O, Ortega E (2010a) Integrated environmental assessment of biodiesel production form soybean in brazil. J Clean Prod 18(1):55–70

Cavalett O, Ortega E (2010) Integrated environmental assessment of biodiesel production from soybean in Brazil. J Cleaner Production 18(1):55–70

Cruz RVA, Nascimento CAO (2012) Emergy analysis of oil production from microalgae. Biomass Bioenerg 47:418–425

Cvengro J, Povaz AF (1996) Production and treatment of rapeseed oil methyl esters as alternative fuels for diesel engines. Biores Technol 55:145–150

Diaz-Chavez RA (2011) Assessing biodiesel: aiming for sustainable development or complying with the market? Energy policy 39(10):5763–5769

Dong XB (2008) Energy and emergy evaluation of bioethanol production from wheat in Henan Province, China. Energy Policy 36(10):3882–3892

Gnansounou E (2011) Assessing the sustainability of biofuels: a logic-based model. Energy 36 (4):2089–2096

Goh CS, Lee KT (2010) Palm-based biofuel refinery (PBR) to substitute petroleum refinery: an energy and emergy assessment. Renew Sustain Energy Rev 14(9):2986–2995

Hattori T, Morita S (2010) Energy crops for sustainable bioethanol production, which, where and how? Plant Production Sci 13:221–234

Hou J, Zhang P, Yuan X, Zheng Y (2011) Life cycle assessment of biodiesel from soybean, jatroha, and microalgae in China conditons. Renewable Sustainable Energy Rev 15(9):5081–5091

Indian Petroleum & Natural Gas Statistics 2012–13, Government of India Ministry of Petroleum & Natural Gas Economic Division New Delhi; Source: Economic Survey, RBI, Ministry of Statistics/Industry/Finance

Jena J, Misra RD (2014) Estimation of production cost of pure plant oils and biodiesels from karanja, palm and soapnut plantations through financial analysis. Small scale forestry

Ju LP, Chen B (2011) Embodied energy and emergy evaluation of a typical biodiesel production chain in china. Ecol Model 222(14):2385–2392

Kaieda M, Samukawa T, Matsumoto T, Ban K, Kond A, Shimada Y (1999) Biodiesel fuel production from plant oil catalyzed by Rhizopus oryzae lipase in a water-containing system without an organic solvent. J Biosci Bioeng 88:627–631

Kim S, Dale BE (2005) Life cycle assessment of various cropping systems utilized for producing biofuels: bioethanol and biodiesel. Biomass Bioenerg 29(6):426–439

Liang Hanwei, Ren Jingzheng, Dong Liang, Gao Zhiuqiu, Zhang Ning, Pan Ming (2016) Is the hydrogen production from biomass technology really sustainable? Answer by life cycle emergy analysis. Int J Hydrogen Energy 41:10507–10514

Liu S, Sun D, Wan S (2007) Emergy evaluation of a kind of biodiesel production system and construction of new emergy indices. J Nanjing Univ (Nat Sci). 43:111–118
Liu J, Lin B-L, Sagisaka M (2012) Sustainability assessment of bioethanol and petroleum fuel production in Japan based on emergy analysis. Energy Policy 44:23–33
Lu H, Lin B-L, Campbell DE, Sagisaka M, Ren H (2012) Biofuel versus biodiversity? Integrated emergy and economic cost-benefit evaluation of rice-ethanol production in Japan. Energy 46:442–450
Matsumura Y, Inoue T, Komoto K, Hirata S, Hada K, Fukuda K (2005) The scale of biomass production in Japan. Biomass Bioenerg 29:321–330
Miao X, Wu Q (2006) Biodiesel production from heterotrophic microalgal oil. Biores Technol 97:841–846
Mohamadabadi HS, Tichkowsky G, Kumar A (2009) Development of a multi-criteria assessment model for ranking of renewable and non-renewable transportation fuel vehicles. Energy 34 (1):112–125
Mu HF, Feng XA, Chu KH (2011) Improved emergy indices for the evaluation of industrial systems incorporating waste management. Ecol Eng 37(2):335–342
Nilsson D (1997) Energy, exergy and emergy analysis of using straw as fuel in district heating plants. Biomass Bioenergy 13(1):63–73
Nimmanterdwong P, Chalermsinsuwan B, Piumsomboon P (2015) Emergy evaluation of biofuels production in Thailand from different feedstocks. Ecol Eng 74:423–437
Nzila C, Dewulf J, Spanjers H, Tuigong D, Kiriamiti H, Langenhove H (2012) Multi criteria sustainability assessment of biogas production in Kenya. Appl Energy 93:496–506
Odum HT (1988) Self-organization, transformity, and information. Science 242(4882):1132–1139
Odum HT (1996) Environmental accounting: emergy and environmental decision making. Wiley
Odum HT (2000) Emergy of global processes, Handbook of emergy evaluation. Gainesville, USA: center of environmental policy. Environmental engineering sciences, University of Florida; Folio 2
Ong HC, Mahlia TMI, Masjuki HH, Honnery D (2012) Life cycle cost and sensitivity analysis of palm biodiesel production. Fuel 98:131–139
Panichelli L, Dauriat A, Gnansounou E (2009) Life cycle assessment of soybean-based biodiesel in Argentina for export. Int J Life Cycle Assess 14(2):144–159
Pereira CLF, Ortega E (2010) Sustainability assessment of large-scale ethanol production from sugarcane. J Clean Prod 18(1):77–82
Ramadhas AS, Jayaraj S, Muraleedharan C (2005) Biodiesel production from high FFA rubber seed oil. Fuel 84:335–340
Saka S, Kusdiana D (2001) Biodiesel fuel from rapeseed oil as prepared in supercritical methanol. Fuel 80:225–231
Scott JA, Ho W, Dey PK (2012) A review of multi-criteria decision-making methods for bioenergy systems. Energy 42:146–156
Selmi B, Thomas D (1998) Immobilized lipase-catalyzed ethanolysis of sunflower oil in solvent-free medium. J Am Oil Chem Soc 75:691–695
Shimada Y, Watanabe Y, Sugihara A, Tominaga Y (2002) Enzymatic alcoholysis for biodiesel fuel production and application of the reaction to oil processing. J Mol Catal B Enzym 17:133–142
Smith CL (2012) The energy challenge. Appl Petrochemicals Res 2:3–6
Spinelli D, Jez S, Pogni R, Basosi R (2013) Environmental and life cycle analysis of a biodiesel production line from sunflower in the Province of Siena (Italy). Energy Policy 59:492–506
Stoeglehner G, Narodoslawsky M (2009) How sustainable are biofuels? Answer and further question arising from an ecological footprint perspective. Biores Technol 100(16):3825–3830
Takahasi F, Ortega E (2010) Assessing the sustainability of Brazilian oleaginous crops-possible raw material to produce biodiesel. Energy Policy 38(5):2446–2454
The National Environmental Accounting Database (NEAD) (2008)
Zhou Z, Jiang H, Qin L (2007) Life cycle sustainabililty assessment of fuels. Fuel 86(1–2):256–263

Chapter 7
Impact of Tri-Fuel on Compression Ignition Engine Emissions: Blends of Waste Frying Oil–Alcohol–Diesel

Thokchom Subhaschandra Singh and Tikendra Nath Verma

Abstract The increasing demand for power and energy pushes mankind in exploiting more on the limited natural resources. This results in the enormous depletion of the overall deposition of natural resources. It has been estimated that the total oil reserve in the world will not last for more than 60 years. Hence, the exploration for alternative fuels is at large. One such alternative fuel for automotive engines is the waste frying oil biodiesel. Waste frying oil methyl esters (WFOMe) have been known to have good engine characteristics as the lower heating value of most methyl esters from waste frying oil ranged from 38.4 to 39.6 MJ/kg, which is close to 44.5 MJ/kg of diesel. Hence, an attempt is being made in reducing further emission of diesel engine and increasing performance and combustion characteristics of the engine with addition of alcohol (10% ethanol) in the blends of diesel and methyl esters of waste frying oil (B20 blend) forming a tri-fuel WFOMe20E10. Locally, sourced materials were collected and characterized for the study and hence prove to be economical while meeting the energy demands.

Keywords Emission · Diesel engine · Waste frying oil · Alcohol
Diesel

7.1 Introduction

There have been various investigations being conducted by researchers across the globe for meeting the crisis of energy demand and reduce the pollution caused by the fossil fuels. This pushes the researchers and scientist onto more environmentally friendly and less polluting sources of biofuels. As diesel engines have been long known to deliver more power and for which it has been used for various commercial and industrial applications, more focus is being driven toward a sustainable

T. Subhaschandra Singh (✉) · T. N. Verma
Department of Mechanical Engineering, National Institute of Technology Manipur,
Langol, Imphal 795004, Manipur, India
e-mail: th.subhas143@gmail.com; th.s.singh@nitmanipur.ac.in

A. K. Agarwal et al. (eds.), *Methanol and the Alternate Fuel Economy*,
Energy, Environment, and Sustainability,
https://doi.org/10.1007/978-981-13-3287-6_7

biofuel, known as biodiesel. Diesel engines have comparatively lower emissions for harmful pollutants like hydrocarbons, carbon monoxide as compared to gasoline engines. Even though other forms of energy sources have been investigated and discovered, biodiesels are highly preferred for reasons such as good oxygen content in fuel (about 15% by weight), stability in storing and ease of transportation. They are also known for non-presence of sulfur and aromatics compounds.

7.1.1 Alternative Fuels

Alternative fuels are those renewable substances which when combusted releases energy as useful work and are normally considered as a substitute for conventional fossil fuels. Though gas fuels have advantages, they pose hazard for conventional engines which are being used for transport and commercial purpose from safety point of view. Hence, biofuels in its liquid state have been considered as another best option for conventional diesel engines. Some of the most commonly used feedstock include waste cooking oil, linseed oil, fish oil, jatropha curcas, cottonseed (Chen et al. 2017; Puhan et al. 2009; Gnanasekaran et al. 2016; Chouhan et al. 2013; Daho et al. 2013), waste plastics (Devaraj et al. 2015; Hurdogan et al. 2017), pongamia seeds, rice bran and neem seeds, koroch seed (Perumal and Ilangkumaran 2017; Ganeshram 2017; Gogoi and Baruah 2011), algae biomass (Vassilev and Vassileva 2016; Satputaley et al. 2016; Rajak and Verma 2018), aeglemarmelos (Krishnamoorthi and Malayalamurthi 2017), chlorella protothecoides (Garcia et al. 2013), coconut oil, palm oil, rapeseed oil, soybean, sunflower oil, cottonseed oil, karanja (Zareh et al. 2017; Tamilselvan et al. 2017; Zaharin et al. 2017; Gharehghani et al. 2017; Ashraful et al. 2014), beef oil, chicken fat, lard oil (Sander et al. 2018; Alptekin et al. 2015; Verma 2018), butanol, ethanol, methanol (Zaharin et al. 2017; Jamrozik 2017; Anand et al. 2010; Verma 2018). Hence, biofuels have an important role in reducing pollution and meeting the energy demands as it is a renewable, non-toxic, biodegradable and has similar characteristics as that of mineral diesel.

Among the sources for biofuel production, waste frying oil has been considered as one of the most economical sources through which sustainable fuel can be produced. The reason, being a waste oil, which is unhygienic for further consumption and is normally disposed-off in large quantities from large hotels/restaurants and even from residential homes. As they do not have specific purpose after being used for few times, frying oil usually ends up in drains, water bodies, and other vegetative fields, eventually causing environmental damage in both land and water. The extractions of methyl esters from waste frying oil include a transesterification reaction through the use of a homogenous/heterogeneous catalyst. Even though there exists a possible source for producing biodiesel, due to its lower cetane number, higher viscosity and density, its direct application in internal combustion engines is not recommended.

One such method through which the properties of biodiesels be improved is through addition of alcohols. Even though addition of alcohols in fuel mixture improves the combustibility of the fuel by increasing viscosity and density while providing oxygen-rich environment, parameters such as lower cetane number, lower value of heating, and longer ignition delay hinders its immense application as fuel. Additions of alcohols in fuel mixture have been reported for lowering emissions from diesel engine but decreases power output of the engine. The reports also suggested increase in brake-specific fuel consumption (BSFC) (Diesel 2011). Among all the reported alcohols, methanol–ethanol can be produced easily in larger quantities. Though methanol is commercially produced from fossil fuel-based products, the degree of miscibility with mineral diesel has been reported to be rather difficult by many researchers (Atmanli 2016; Yao et al. 2008). Ethanol, on the other hand, has good miscibility features with that of petroleum-based fuels. Due to its higher oxygen content, it improves the overall performance of the blended fuel. It is also being reported that blending ethanol to diesel decreases pollutants from engine emissions (Emiro and Mehmet 2018). Ease of production from biomass, such as rice straw, rice husks, corn, sugar cane carcass (straw/leaves), barley through the process of fermentation, has been an important characteristic of ethanol. Hence, ethanol is being an interesting choice by many researchers for its abundance and better oxygen content.

7.1.2 Recent Advancements in Internal Combustion Engines with Biofuels

Biofuels have specific properties and have the tendency for reduction of emissions, as compared to fossil fuels, from internal combustion engines. Though biofuels have enormous properties which are close to that of fossil fuels, direct application of biofuels in the existing internal combustion engines is not recommended. This is primarily due to lower calorific value, high density, low cetane number, poor auto-ignition characteristics, etc. Thus, various methods are being applied and tested by numerous researchers in the area of internal combustion engineering. Mixtures of alcohols (ethanol–methanol–butanol) with cottonseed biodiesel and diesel have been tested for its combustion, performance, and emission characteristics. It was reported that the ignition delay was longer in the case of blended fuel due to their lower cetane number, as compared to diesel. Values of maximum cylinder pressure and maximum heat release rate were observed to lie close to that of diesel, with increase in BSFC (Emiro and Mehmet 2018). Another study uses biodiesel with ethanol–butanol blends on an engine and found that the blended fuel has adverse effects on the performance of the engine. It also increases the emission of carbon monoxide and hydrocarbons by about 13–22% and 5–29%, respectively, yet it decreases oxides of nitrogen by 6–28% for biodiesel–ethanol and biodiesel–butanol blends, respectively. They have concluded that ethanol has more emission reduction capability than that of butanol (Wei et al. 2018). A study by Soto et al. (2018) has shown that B20 blends of diesel and biodiesel have higher engine characteristics, compared to diesel. They have reported increase in power by 1.2, 1% in torque, 1.2% in brake thermal efficiency (BTE), while reduction in soot concentrations by about 8.9% (Soto et al. 2018). Other studies include fumigation of gasoline on B20 blend on a diesel engine. The results show subsequent decrease in exhaust gas temperature (EGT), oxides of nitrogen, and carbon dioxide, which increasing (BTE) (Hoseinpour et al. 2018).

The addition of nano-additives, such as manganese and cobalt on B20 biodiesel of waste frying oil and diesel, has significantly improved the BSFC and BTE while reducing oxides of nitrogen and carbon monoxide (Mehregan and Moghiman 2018). A study by Aldhaidhawi et al. (2017) has reported the effect of B20 blend biodiesel with hydrogen (0–5%) on the engine performance, combustion, and mixture forming characteristics. It was observed that emissions such as carbon monoxide, smoke, and unburned hydrocarbons have been reduced, while an increase in oxides of nitrogen is observed at all engine speeds (Aldhaidhawi et al. 2017). Nanoparticles of alumina as additive to the B20 biodiesel produced from poultry litter oil have been investigated in a diesel engine. The study reported that B20 has improved the performance and combustion characteristics, with and without additives. There has been substantial decrease in emissions of carbon monoxide (CO), unburned hydrocarbons (HC), and oxides of nitrogen (NO) of the B20 blended biodiesel with alumina nanoparticles additives (Ramesh et al. 2018). Other additives such as di-*tert*-butyl have been suggested for use with palm oil methyl ester–diesel blended B20 biofuel in a compression ignition engine. The study revealed improvements in BTE by about 2–3.5% and reduction in specific energy consumptions by 10–15%. The absence of additives in the B20 blend also results in lower CO and NO_x while slight increase in unburned HC has been observed (Musthafa et al. 2018). A performance and emission characteristics study by Wu et al. (2018) using biodiesel–diesel blend with ethanol of 4% in B10 in the presence of carbon-coated aluminum nanoparticles as additives show increase in total HC emission due to the presence of ethanol in the fuel blend. Addition of the nanoparticles has shown reduction of BSFC by about 6%, NO_x by about 6%, and CO by about 19%. The study suggested that after the combustion phase the carbon-coated aluminum nanoparticles were converted into alumina nanoparticles (Wu et al. 2018). Other studies using ferrous-based nanoparticles suggested reduction in emissions while increasing the efficiency of the cycle (Kumar et al. 2017). Parametric studies involving different engine parameters of diverse blend ratios for various biodiesels such as waste cooking oil, bauhinia variegata, neem, jatropha, hybrid pongamia have suggested that the use of biodiesel in engine at specific blend ratio of B20 has tendency of reducing emissions without much compromise in performance and combustion characteristics (How et al. 2018; Abed et al. 2018; Yatish et al. 2018; Nair et al. 2017; Patil et al. 2018; Perumal and

Ilangkumaran 2018; Suresh et al. 2018; Rajak and Verma 2018). Hence, it is evident that the use of alcohols has the tendency in reducing emissions. Though various studies suggested the applicability of biofuels and/or alcohol in an engine, the combined effect investigations are by far still the least explored.

7.1.3 *Emission Norms of India*

The Government of India (GoI) has laid down specific rules and regulations which must be adhered by the vehicle manufacturers so as to limit the level of pollutions from vehicular emissions, based on the year and type of implementation. The reference for the formulations has been taken with reference to the EURO norms. The regulations in India have been referred to as India 2000 for EURO-1 and thereafter as Bharat Stage (BS) from EURO-2 onwards. Table 7.1 shows the Indian emissions standards for four-wheeled vehicles and Table 7.2 shows the standards followed with respect to time frame (https://www.dieselnet.com/standards/in/). Commercial vehicles have also posed a great hazard in the transportation sector. This may be due to the improper maintenance and irregular check-up of the commercial transport vehicles. Diesel engines in general have been known to emit high amounts of pollution from its engine exhaust (Singh et al. 2018).

From the literature, it is evident that waste frying oil can be a suitable alternative to be blended with diesel, for producing sustainable biofuels. Yet researchers have suggested the use of B20 blend, since it has similar characteristics to that of diesel. Among the additives, ethanol has been selected as it can be easily produced from biomass and production is easy. Still, more amounts of ethanol blend fuel result in emission of other unwanted compounds from the engine. Hence, the authors are making an attempt to bring out a suitable fuel by using tri-fuel, a mixture of waste frying oil, ethanol and diesel on specific blend ratio of 20% waste frying oil, with 10% ethanol in B20 blend on a single cylinder, naturally aspirated, compression ignition engine, for studying the emission characteristics of the produced fuel at variable loading conditions.

Table 7.1 Indian emission standards

S. No.	Reference	Indian standard	Year of nationwide implementation
1	EURO-1	India 2000	2000
2	EURO-2	BS-II	2001–05
3	EURO-3	BS-III	2005–10
4	EURO-4	BS-IV	2010–17
5	EURO-5	BS-V	NA
6	EURO-6	BS-VI	2020

Table 7.2 Indian emission standards for various vehicle segments (https://www.dieselnet.com/standards/in/)

S. No.	Category	Type	Stage	Year	Permissible emission levels (g/km)			
					CO	HC	NO_x	PM
1	Passenger cars	Gasoline		1991	14.3	2.0	–	–
				1996	8.68	–	–	–
				1998	4.34	–	–	–
			India 2000	2000	2.72	–	–	–
			BS-II	2005	2.2	–	–	–
			BS-III	2010	2.3	0.20	0.15	–
			BS-IV	2017	1.0	0.10	0.08	–
			BS-VI	2020	1.0	0.068	0.06	–
		Diesel		1992	17.3	2.7	–	–
				1996	5.0	–	–	–
			India 2000	2000	2.72	–	–	0.14
			BS-II	2005	1.0	–		0.08
			BS-III	2010	0.64	–	0.50	0.05
			BS-IV	2017	0.50	–	0.25	0.025
			BS-VI	2020	0.50	–	0.080	0.0045
2	Light duty	Diesel		1992	14.0	3.5	18.0	–
				1996	11.20	2.40	14.4	–
			India 2000	2000	4.5	1.1	8.0	–
			BS-II	2005	4.0	1.1	7.0	–
3	Heavy duty	Diesel		1992	17.3	2.7	–	–
				1996	11.2	2.4	14.4	–
			India 2000	2000	4.5	1.1	8.0	0.36
			BS-II	2005	4.0	1.1	7.0	0.15
			BS-III	2010	2.1	0.66	5.0	0.10
			BS-IV	2010	1.5	0.46	3.5	0.02
			BS-V	NA	1.5	0.46	2.0	0.02
			BS-VI	2020	1.5	0.13	0.40	0.01
4	Two-wheeler	Gasoline		1991	12–30	8–12	–	–
				1996	4.50	–	–	–
			BS-I	2000	2.00	–	–	–
			BS-II	2005	1.5	–	–	–
			BS-III	2010	1.0	–	–	–
			BS-VI	2020	1.0	0.10	0.06	0.0045

(continued)

Table 7.2 (continued)

S. No.	Category	Type	Stage	Year	Permissible emission levels (g/km)			
					CO	HC	NO_x	PM
5	Three-wheeler	Gasoline		1991	12–30	8–12	–	–
				1996	6.75	–	–	–
			BS-I	2000	4.0	–	–	–
			BS-II	2005	2.25	–	–	–
			BS-III	2010	1.25	–	–	–
			BS-VI	2020	0.44	0.35	0.085	–
6	Three-wheeler	Diesel	BS-II	2005	1.0	–	–	0.10
			BS-III	2010	0.50	–	–	0.05
			BS-VI	2020	0.22	0.10	0.10	0.025

7.2 Materials and Method

The method of fuel preparation comprises of collection of feed materials, synthesis of the methyl esters from waste frying oil in the presence of heterogeneous catalyst, and then proportionate with diesel to produce B20 blend. Ethanol is mixed at 10% to the B20 blend, and the final fuel is prepared.

7.2.1 Feed Sources and Production of Methyl Esters

The waste frying oil (WFO) is collected from nearby hotels, in and around Langol, Imphal West, Manipur. The heterogeneous catalyst is produced from leftover eggshells, which are also abundantly available around the area. The WFO consists of carbon particles in various sizes and has pungent smell. They are preheated to about 40 °C, stirred occasionally, and then let to rest for about 30 min. This process helps in collecting the carbon particles in the bottom of the pan. Using a fine cloth filter of mesh size 0.5 mm, the oil is filtered into a clean beaker. Simultaneously, the eggshells are also washed and pre-dried in an oven for about 2 h at 120 °C. The pre-treated eggshells are then exposed at 900 °C in a muffle furnace for about 6 h to complete the process of calcinations. The calcinated shells are then finely crushed and analyzed of its properties through a scanning electron microscope (SEM), X-ray diffractometer (XRD), and a Fourier-transform infrared spectroscopy (FTIR).

The synthesized calcium oxide has been used for the transesterification process of the WFO for extracting methyl esters. The reaction takes place in a magnetic stirrer coupled heated plate for about 3 h at 65 °C with alcohol to oil ratio of 10 and 5% (wt%) catalyst loading for a batch size of 2 l. The mixture is then placed in an inverted conical beaker for removing glycerol from the mixture. The methyl esters

(WFOMe) were then extracted from the beaker, and the conversion rate was found to be about 94%, using the formula in Eq. (7.1) (Dhawane et al. 2016).

$$\text{WFO FAME yield} = \frac{\text{Weight of WFO FAME produced}}{\text{Weight of WFO procured}} \times 100 \qquad (7.1)$$

The SEM image of the calcinated eggshells at 900 °C in a muffle furnace for about 6 h is shown in Fig. 7.1a. The EDX result shows the presence of calcium and oxides in the sample as shown in Fig. 7.1b. The analysis illustrates weight % and atomic % of Ca as 66.78% and 42.76% while that of O as 25.82 and 41.41%, respectively, therefore confirming CaO presence of about 91 ± 8%. XRD pattern and IR results of the calcinated samples are shown in Fig. 7.1c, d. From the respective figures of XRD and IR, the results are in good relationship with SEM + EDX results, showing the formation of calcium oxide. The CaO was displayed in the bands of 895, 1056, and 1493 cm^{-1} from the IR spectra.

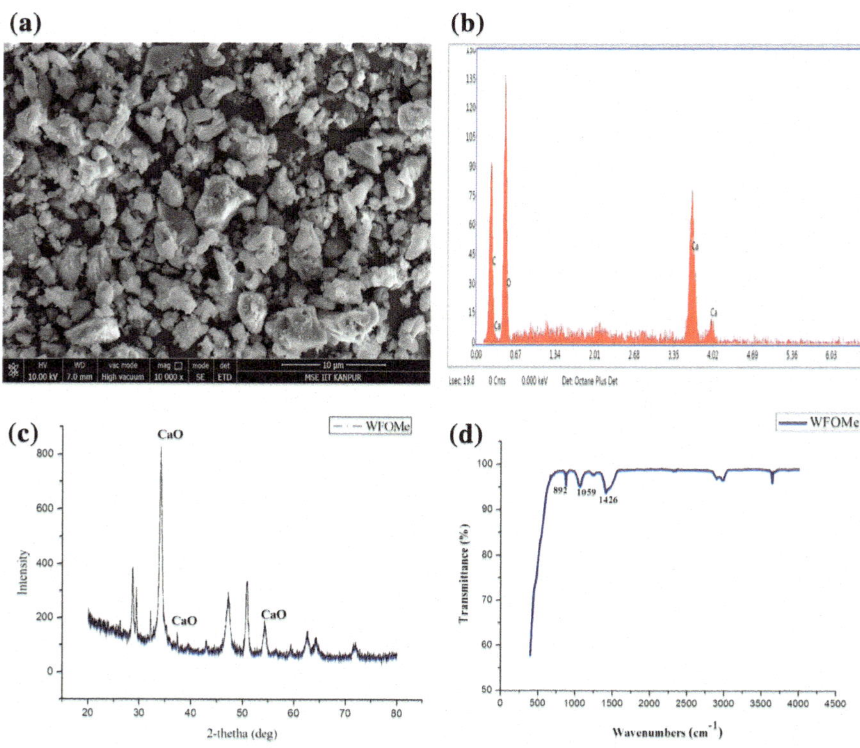

Fig. 7.1 Characterization of CaO catalyst using **a** SEM, **b** EDX, **c** XRD, and **d** FTIR

7.2.2 *Preparation of Blends and Engine*

The WFOMe were then taken into a clean beaker and rested. The sample is then rested for about 10 days and to investigate any leftover particles in the synthesized oil. Thin layer chromatography (TLC) test was conducted to ensure any unwanted compounds from the sample. Then a conical flask is taken, and proportionate 20% is taken from the WFOMe beaker and mixed with 80% diesel. The mixture is then stirred and then kept for about 30 days to study the stability of the mixture, renamed as WFOMe20. At the end of the stipulated time, satisfactory results have been observed from the fuel mixture. Then, locally sourced ethanol from the market produced from locally available raw materials in and around Imphal, India, has been procured. The ethanol is then transferred into a clean beaker, from which about 10% is added to the WFOMe20, thus forming WFOMe20E10. The fuel properties are provided in Table 7.3. The density of the fuel is measured at 20 °C using ASTM standard D4052. The kinematic viscosity of the fuel on the other hand is being measured using ASTM D445 at 40 °C. An IKA C200 bomb calorimeter with gelatin capsules is used for determining the heating value of fuels. The cetane number is measured using the four variable equations, conforming to ASTM D4737 through the use of a density meter at 20 °C. The flash point is determined by open-cup apparatus. Figure 7.2 shows the blended fuels in the laboratory of chemistry department.

The tests are being carried out under variable loading conditions of no-load, half load, and full load conditions, with constant engine speed of 1500 rpm. The test of the fuels, both diesel, WFO, WFOMe20, and WFOMe20E10, were carried out under similar operating conditions to ensure comparison with reference diesel. The experimental setup is being shown in Fig. 7.3. The engine experiment is carried out on a single cylinder, naturally aspirated, eddy current dynamometer, water cooled, and direct injection diesel engine. The properties of the test engine are shown in Table 7.4. The engine is coupled with K-type thermocouples for measuring temperatures at various locations such as inlet manifold, exhaust manifold, cylinder chamber, cooling water circuit, calorimeter inlet, calorimeter outlet.

Table 7.3 Properties of standard, synthesized, and blended fuel

S. No.	Parameter	Diesel	WFO	WFOMe20	WFOMe20E10
1	Density (kg/m^3)	831.5	883	851	846
2	Kinematic viscosity @40 °C (mm^2/s)	2.8	3.2	2.96	2.74
3	Lower heating value (MJ/kg)	42.5	38.3	41.7	39.3
4	Cetane number	58.6	53.2	57.3	48.6
5	Flash point (°C)	72	137	64	61

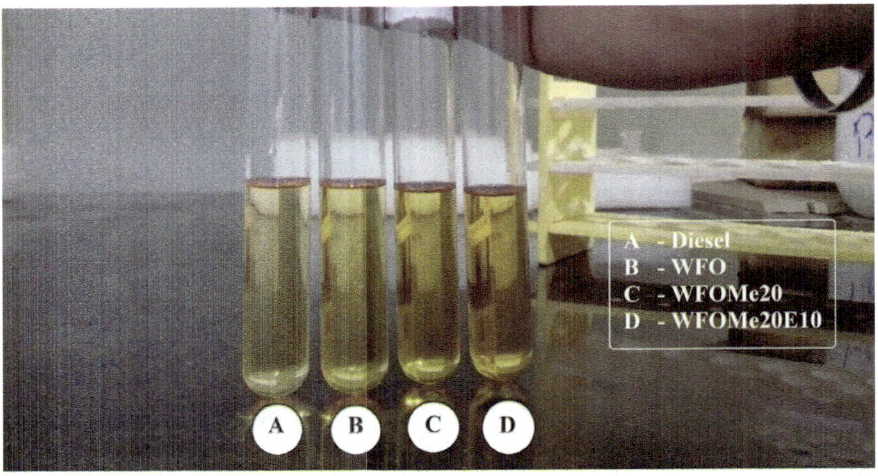

Fig. 7.2 Surrogate fuels and diesel

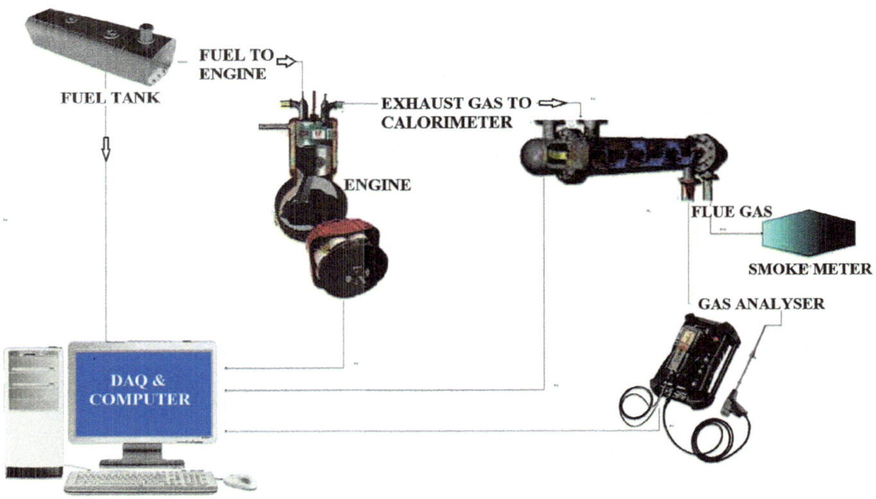

Fig. 7.3 Experimental setup

7.2.3 *Uncertainty Analysis*

In every experimental measurement, there is always a chance of creeping in errors from various sources during the course of the experiment. These errors are normally of random nature while some of them are from the gross howler of the researcher/

Table 7.4 Properties of test engine

S. No.	Parameter	Specification
1	Make and model	Legion brothers; MFVCR
2	Cylinder	1
3	Cooling	Water
4	Compression ratio	17.5
5	Rated speed	1500 rpm
6	Bore × stroke	80 × 110
7	Fuel injection pressure	220 bar
8	Starting	Manual cranking
9	Fuel delivery (°CA)	23 deg BTDC
10	Maximum power	5 HP/3000 rpm
11	Nozzle	3 holes/multihole/120 deg
12	Intake valve (°CA)	4.5 deg BTDC/35.5 deg ATDC

experimenter. Even though bad data may just be neglected, many data may just seem bad and are not actually bad at all. For instance, some of the errors may be discovered after repetition of the experiment, thereby finding the uncertainty of the experiment. Yet there are certain experiments where there is involvement of huge cost and this will hinder multiple experiments. Hence, researchers must take extra care to ensure that final result from the primary readings have minimum errors or uncertainty. If w_1, w_2, $w_{3....}$ $w_{n,}$ be the uncertainties of the individual parameters, such as pressure sensor, temperature sensor error, then the total uncertainty of the experiment w_R can be calculated using Eq. (7.2) (Holman 2012). The uncertainty of individual parameters for the experiment is shown in Table 7.5. The uncertainty of

Table 7.5 Uncertainties of parameters

S. No.	Instrument	Uncertainty (%)
1	Pressure sensor	±0.5
2	Temperature sensor	±0.15
3	Load sensor	±0.2
4	Fuel sensor	±0.5
5	Crank angle encoder	±0.2
6	Speed sensor	±1.0
7	Smoke meter	±1.0
8	Testo 350 flue gas analyser	
	CO	±0.3
	CO_2	±1.0
	HC	±0.1
	NO_x	±0.5
	O_2	±0.3

measuring parameters for performance and combustion characteristics is ±1.265%, while that of emission parameters is ±1.562%.

$$\text{Uncertainty } \%(w_R) = \sqrt{\left(\frac{\partial R}{\partial x_1} w_1\right)^2 + \left(\frac{\partial R}{\partial x_1} w_1\right)^2 + \left(\frac{\partial R}{\partial x_1} w_1\right)^2 + \cdots \left(\frac{\partial R}{\partial x_n} w_n\right)^2} \quad (7.2)$$

7.3 Results and Discussion

7.3.1 Performance and Combustion Characteristics

The performance and combustion characteristics of the tested fuels are discussed herein the section. The brake thermal efficiency (BTE) of diesel, surrogate fuel with and without additive at constant engine speed of 1500 is shown in Fig. 7.4. The BTE is a function of calorific value of the fuel, and hence, diesel being the highest among all the tested fuels has the highest BTE at all engine loads. Though WFOMe and ethanol have lesser lower heating value than diesel, surrogate fuels such as WFOMe20 and WFOMe20E10 have been showing similar trends like diesel in BTE of the test engine, at all loads. The values of BTE have increased with increase in engine loading for all fuels, while results of WFOMe20 and WFOMe20E10 have been seen to have close value of BTE at full load conditions. Similarly, lower heating values have also resulted in higher brake-specific fuel consumption (BSFC) by the test engine since more fuel is required during injection to obtain similar power output. The BSFC for the tested fuels at various loads is shown in Fig. 7.5. The BSFC for all fuels has decreased significantly with increase in engine loading till partial load, but increases slightly as the loading reaches full

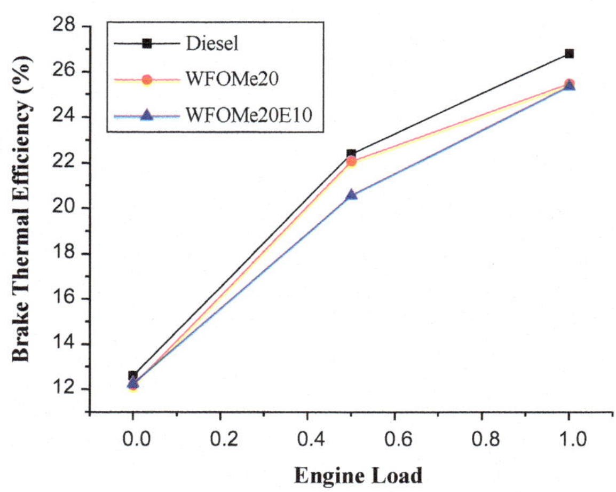

Fig. 7.4 Comparison of BTE against engine load for all tested fuels

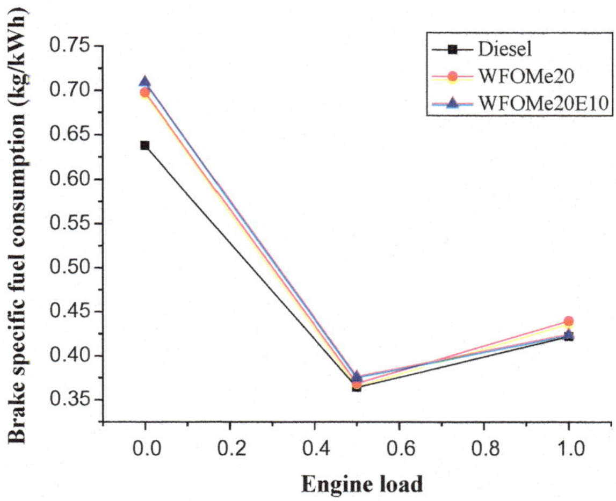

Fig. 7.5 Comparison of BSFC against engine load for all tested fuels

load condition. This may be due to lower premixed phase during combustion (Emiro and Mehmet 2018).

Combustion characteristics such as cylinder pressure, heat release, and exhaust gas temperatures are discussed herein. The early start of injection is primarily due to high viscosity and low compressibility of fuels inside an engine chamber by rapidly raising the fuel pressure. The injection of fuel is earlier by about 2–3%, as compared to diesel, for surrogate fuels using biofuel and alcohol additives (Emiro and Mehmet 2018).

Figure 7.6 shows the variation of cylinder pressure with crank angle for no-load conditions. From the figure, it can be seen that WFOMe20 and WFOMe20E10 have higher cylinder pressure as compared to diesel. This is primarily due to more injection of fuel in the combustion chamber, due to lower heating value of the fuels and cetane number, in producing equivalent output, with that of diesel. The peak value of cylinder pressure for the surrogate fuels lies close to that diesel. Figure 7.7 shows the heat release rate (HRR) of all tested fuels at no-load condition with engine operating at constant 1500 rpm speed. The HRR is another important parameter in understanding the advancement of combustion for all tested fuels in the engine chamber. From the illustration, it can be noted that the HRR for WFOMe20 has lower values than that of diesel. Yet, the surrogate fuel with ethanol additive, WFOMe20E10, is seen to have higher values of HRR, as compared to diesel and WFOMe20. This study is also supported by other researchers using ethanol as additives (Emiro and Mehmet 2018; Wei et al. 2018).

While the variation of exhaust gas temperature (EGT) with engine load for all tested fuel on Fig. 7.8 shows that, the EGT of all tested fuels increases with increase in engine loading. Due to more oxygen content in the surrogate fuels, lower calorific value, and less cetane number, a higher amount of fuel are being injected and this increases the temperature inside the combustion chamber. This

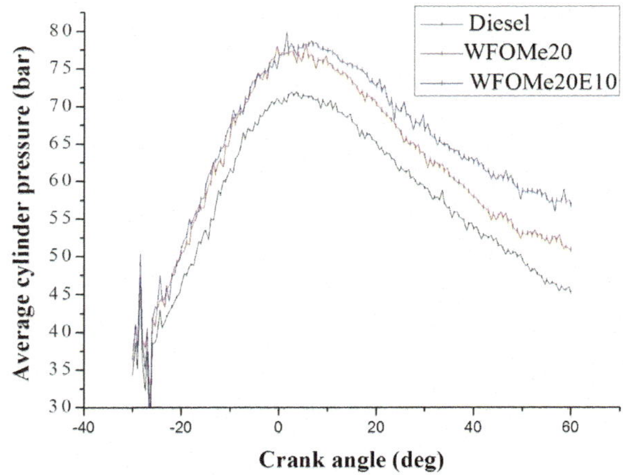

Fig. 7.6 Variation of cylinder pressure with CA (deg)

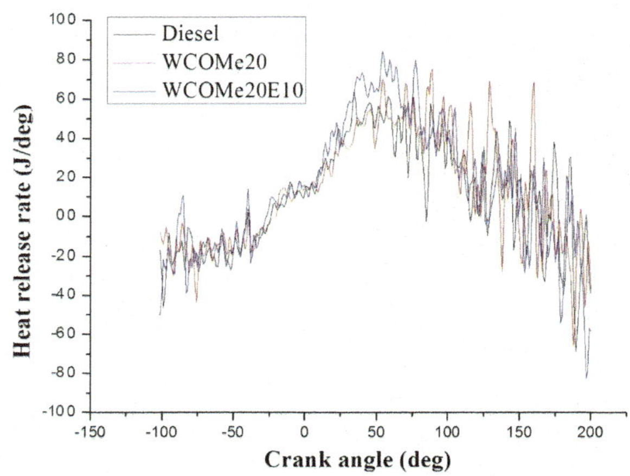

Fig. 7.7 Variation of heat release rate with CA (deg)

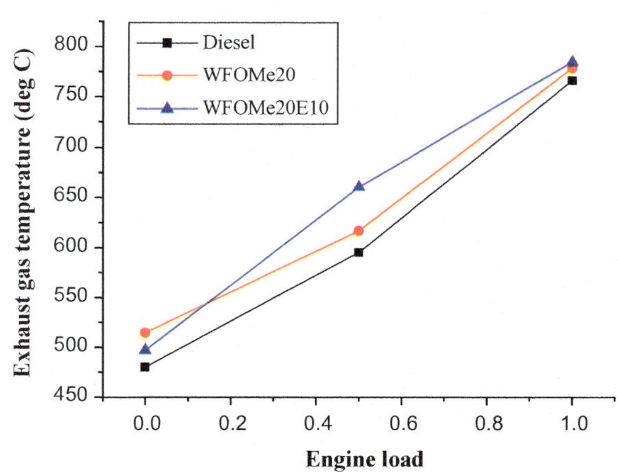

Fig. 7.8 Comparison of EGT against engine load for all tested fuels

results in higher EGT of the surrogate fuel than diesel. On the contrary to emissions, higher EGT also depicts better combustibility inside the combustion chamber.

7.3.2 *Emission Characteristics*

Emission characteristics are an important parameter in understanding engine parameters. Also, diesel engines are known to emit harmful pollutants which are categorized in the Group-I carcinogens. Hence, when using biofuels and other additives, the need to study emissions from the engine is obligatory. Though diesel engines in general have various chemicals as pollutants from exhaust emissions, the exhaust gas analyzer has limited the study into few parameters only (Singh et al. 2018). One among them is the hydrocarbons (HC) emission. They are caused by many factors, such as crankcase emission, rich supply of fuel at lower temperatures, flame quenching at the walls during expansion stroke. The total HC emissions for the surrogate fuels with respect to diesel for various engine loading is shown in Fig. 7.9. The use of WFOMe20 as a fuel has close characteristics to that of diesel at partial loads, while at full load conditions, the total HC have increased. WFOMe20E10 have less effect in reducing total HC when uses as surrogate fuel in diesel engine at all loads.

Carbon monoxide (CO) emissions are due to the incomplete combustion inside the engine chamber. In order to understand the combustibility of a fuel in the engine chamber, CO emission study is an inevitable factor. The formation of CO during combustion for all tested fuels in the test engine is shown in Fig. 7.10. Rich mixture during fuel injection is another main cause for incomplete burning, and hence, operating in bi-fuel mode reduces the emission of CO. As depicted in the illustration, WFOMe20 have lesser CO emission than diesel. Addition of ethanol in the surrogate fuel of WFOMe20 by 10% has increased the combustibility of the

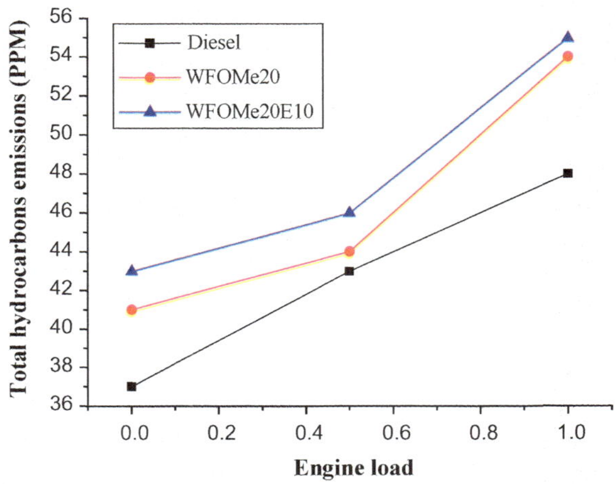

Fig. 7.9 Variation in total HC emissions

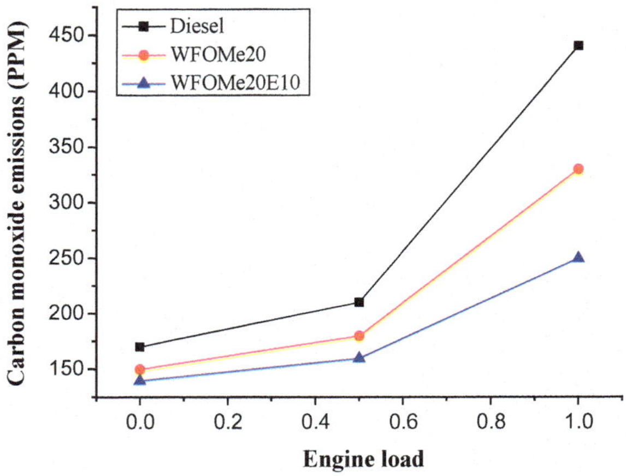

Fig. 7.10 Variation in CO emissions

mixture since ethanol is known to increase oxygen content in the fuel. Thus, lesser CO emission is observed from the WFOMe20E10 fuel at all loads, as compared to diesel and WFOMe20. During the increase in engine loading, scarcity of oxygen in the combustion chamber has resulted in increase of CO emission for all fuels.

Though oxides of nitrogen have been known to be produced in lesser quantities in various sectors like industries and at elevated temperature areas, continuous exposure to it causes harm to the environment and living beings. They have been known to cause asthma, bronchitis, etc., in humans. NO_x is generally formed due to chemical bonding of atmospheric nitrogen and oxygen at elevated temperatures. Such a favorable condition is available inside the combustion chamber of compression ignition engines (Singh et al. 2018). The emissions of oxides of nitrogen for the tested fuels in all engine loads are shown in Fig. 7.11. As observed from the illustration, the percentage of NO_x in PPM at no-load condition is almost equal for

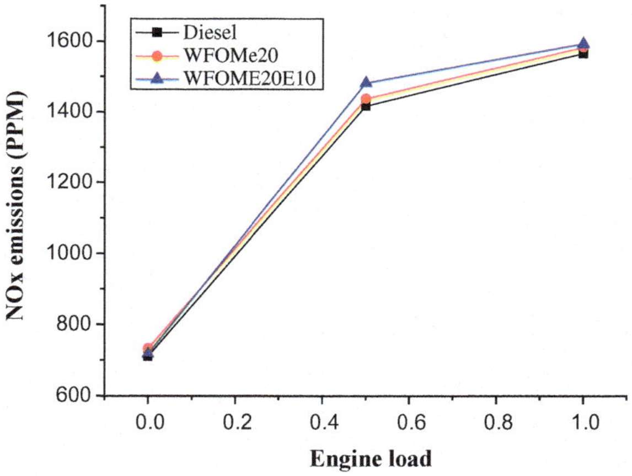

Fig. 7.11 Variation in NO_x emissions

all tested fuels. But as the engine loading increases, the amount of fuel injected inside the combustion chamber increases and more heat is being produced, thereby elevating the temperature inside the chamber. This causes the nitrogen to easily combine with oxygen from the fuel in producing NO_x in the engine exhaust. The amount of NO_x for WFOMe20E10 comes close to that of diesel at full loading condition with approximately 1400 PPM of NO_x for all tested fuels.

Though carbon dioxide (CO_2) emission is considered to be a good sign for complete combustion, emission at larger quantity not only increases the global temperature, but also causes harmful damage to the environment and living beings. Properly studying the CO_2 emission from an engine also helps in understanding the combustion characteristics of an engine. The variation of CO_2 emissions with engine load for all tested fuels in the experiment engine is shown in Fig. 7.12. The percentage of CO_2 emission for WFOMe20E10 has been found to be highest followed by WFOMe20. This may be due to the extra supply of oxygen from the methyl esters of waste frying oil and additives. Similar characteristics with that of CO emission can be seen in the CO_2 emissions also, as the engine loading increases. At full load condition, WFOMe20E10 have less emission than WFOMe20. Thus, addition of ethanol as fuel additive and bi-fuel mode increases the combustibility of diesel fuel inside the combustion chamber.

Soot formations by the test engine for all tested fuels are shown in Fig. 7.13. Soot is formed as a result of thick smoke being formed as a by-product of combustion. The additions of biofuel and additives such as ethanol have improved the smoke opacity of the exhaust gas. This results in low formation of soot in the engine exhaust. For all tested engine loads, the soot formation of surrogate fuels WFOMe20 and WFOMe20E10 has lesser values than that of diesel. This not only reduces presence of carbon particles, but also helps in reducing particulate matter (PM) from the engine exhaust.

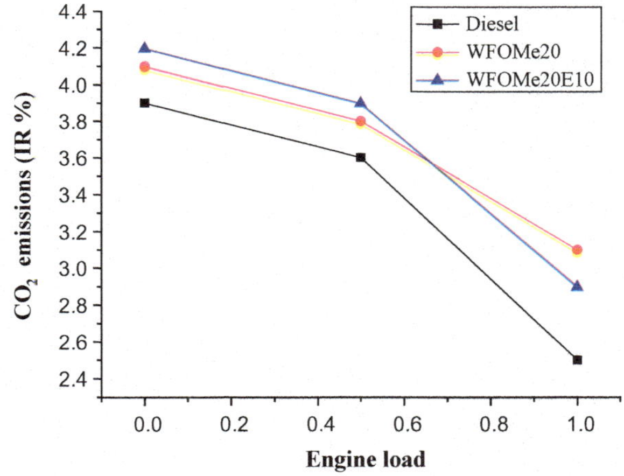

Fig. 7.12 Variation in CO_2 emissions

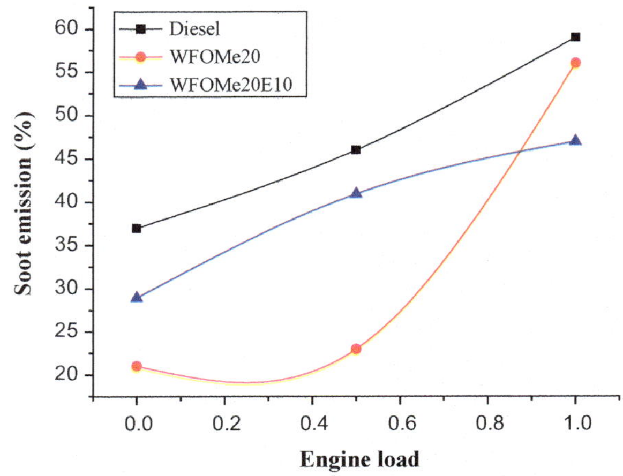

Fig. 7.13 Variation in soot emissions

7.3.3 Characteristic Comparison of Present Study with Past Researchers

A comparative study is also conducted where the exhaust emissions of other researcher's work have been compared with the present study. The comparison is shown in Fig. 7.14. Studies by Aldhaidhawi et al. (2017), Emiro and Mehmet (2018), and Hoseinpour et al. (2018) are being compared for NO_x formation with the present study. It is evident from the graph that the proposed tri-fuel (BF1) in the present study has shown to have lower emissions than those of other biofuels tested by past researchers, except for hydrogen injection with biodiesel.

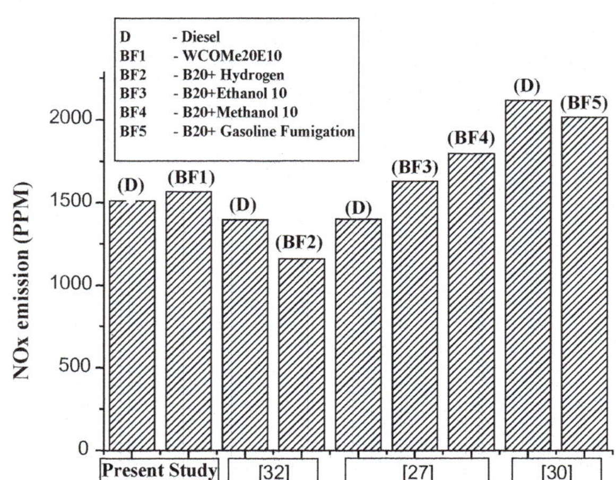

Fig. 7.14 Comparison of NO_x formation of the present study with past researchers

7.4 Conclusion

Locally available resources such as waste frying oil and waste eggshells have been utilized in producing methyl esters. Locally sourced commercial alcohol (ethanol) is being used as an additive in studying the effect of the tri-fuel in engine emissions, performance, and combustion characteristics. Though the calorific values of biofuel and additives are lower than that of diesel, the combination as tri-fuel improves the combustibility of the overall surrogate fuel. The observations from the study are given below:

- Slight decrease in BTE and more BSFC is observed for both WFOMe20 and WFOMe20E10 fuels.
- Peak cylinder pressure values and heat release rate for WFOMe20 and WFOMe20E10 lie close to that of diesel. Maximum HRR is obtained for WFOMe20E10 at no-load condition with engine speed at 1500 rpm.
- Higher EGT is observed for the surrogate fuels, and hence, more NO_x is emitted from the engine, as compared to that of diesel.
- CO_2 emission is more for surrogate fuels, and hence, it shows good combustion of the fuels inside the combustion chamber.
- Soot formation is significantly reduced with use of surrogate fuels, than diesel, showing a possible reduction of particulate matter.

Thus, locally sourcing available waste materials for producing tri-fuel have been proven to be economical and provide a possible energy source for meeting the energy crisis in and around the northeastern province of India.

Acknowledgements The authors sincerely thank Department of Chemistry, NIT Manipur, Imphal, India, for extensive support during the course of the experimental procedure and providing facility such as XRD, IR. The authors also acknowledge IIT Kanpur for providing SEM with EDX facility.

References

Abed KA, El Morsi AK, Sayed MM, El Shaib AA, Gad MS (2018) Effect of waste cooking-oil biodiesel on performance and exhaust emissions of a diesel engine. Egypt J Pet (in Press Corrected proof)

Aldhaidhawi M, Chiriac R, Badescu V, Descombes G, Podevin P (2017) Investigation on the mixture formation, combustion characteristics and performance of a Diesel engine fueled with Diesel, Biodiesel B20 and hydrogen addition. Int J Hydrogen Energy 42:16793–16807

Alptekin E, Canakci M, Necati A, Turkcan A, Sanli H (2015) Using waste animal fat based biodiesels—bioethanol—diesel fuel blends in a DI diesel engine. Fuel 157:245–254. https://doi.org/10.1016/j.fuel.2015.04.067

Anand K, Sharma RP, Mehta PS (2010) Experimental investigations on combustion, performance and emissions characteristics of neat karanji biodiesel and its methanol blend in a diesel engine. Biomass Bioenergy 35:533–41. http://dx.doi.org/10.1016/j.biombioe.2010.10.005

Ashraful AM, Masjuki HH, Kalam MA, Fattah IMR, Imtenan S, Shahir SA, Mobarak HM (2014) Production and comparison of fuel properties, engine performance and emis- sion characteristics of biodiesel from various non-edible vegetable oils: a review. Energy Convers Manage 80:202–228. http://dx.doi.org/10.1016/j.enconman.2014.01.037

Atmanli A (2016) Comparative analyses of diesel—waste oil biodiesel and propanol, n-butanol or 1-pentanol blends in a diesel engine. Fuel 176:209–215

Chen CY, Lee WJ, Wang LC, Chang YC, Yang HH, Young LH et al (2017) Impact of high soot-loaded and regenerated diesel particulate filters on the emission of persistent organic pollutants from a diesel engine fueled with waste cooking oil-based biodiesel. Appl Energy 191:35–43. https://doi.org/10.1016/j.apenergy.2017.01.046

Chouhan APS, Singh N, Sarma AK (2013) A comparative analysis of kinetic parameters from TGDTA of Jatropha curcas oil, biodiesel, petroleum diesel and B50 using different methods. Fuel 109:217–224. https://doi.org/10.1016/j.fuel.2012.12.059

Daho T, Vaitilingom G, Ouiminga SK, Piriou B, Zongo AS, Ouoba S et al (2013) Influence of engine load and fuel droplet size on performance of a CI engine fueled with cottonseed oil and its blends with diesel fuel. Appl Energy 111:1046–1053. https://doi.org/10.1016/j.apenergy.2013.05.059

Devaraj J, Robinson Y, Ganapathi P (2015) Experimental investigation of performance, emission and combustion characteristics of waste plastic pyrolysis oil blended with diethyl ether used as fuel for diesel engine. Energy 85:304–309. https://doi.org/10.1016/j.energy.2015.03.075

Dhawane SH, Bora AP, Kumar T, Halder G (2016) Parametric optimization of biodiesel synthesis from rubber seed oil using iron doped carbon catalyst by Taguchi approach. Renew Energy

Diesel I (2011) The influence of n-butanol/ diesel fuel blends utilization on a small diesel engine performance and emissions. Fuel 90:2467–2472. https://doi.org/10.1016/j.fuel.2011.02.033

Emiro AO, Mehmet Ş (2018) Combustion, performance and exhaust emission characterizations of a diesel engine operating with a ternary blend (alcohol-biodiesel-diesel fuel). Appl Thermal Engg 133:371–380

Ganeshram AK (2017) A comparative study of combustion, emission and performance characteristics of rice-bran-, neem-, and cottonseed-oil biodiesels with varying degree of unsaturation. Fuel 187:296–305. https://doi.org/10.1016/j.fuel.2016.09.062

Garcia MCC, Sanchez MDM, Miron AS, Camacho FG, Grima EM (2013) A process for biodiesel production involving the heterotrophic fermentation of Chlorella protothecoides with glycerol as the carbon source. Appl Energy 103:341–349. https://doi.org/10.1016/j.apenergy.2012.09.054

Gharehghani A, Mirsalim M, Hosseini R (2017) Effects of waste fish oil biodiesel on diesel engine combustion characteristics and emission. Renew Energy 101:930–936. https://doi.org/10.1016/j.renene.2016.09.045

Gnanasekaran S, Saravanan N, Ilangkumaran M (2016) Influence of injection timing on performance, emission and combustion characteristics of a DI diesel engine running on fish oil biodiesel. Energy 116:1218–1229. https://doi.org/10.1016/j.energy.2016.10.039

Gogoi TK, Baruah DC (2011) The use of Koroch seed oil methyl ester blends as fuel in a diesel engine. Appl Energy 88:2713–2725. https://doi.org/10.1016/j.apenergy.2011.02.023

https://www.dieselnet.com/standards/in/

Holman JP (2012) Experimental method for enginers, 8th edn. Mcgraw-hill. ISBN: 978-0073529301

Hoseinpour M, Sadrnia H, Tabasizadeh M, Ghobadian B (2018) Evaluation of the effect of gasoline fumigation on performance and emission characteristics of a diesel engine fueled with B20 using an experimental investigation and TOPSIS method. Fuel 223:277–285

How HG, Masjuki HH, Kalam MA, Teoh YH (2018) Influence of injection timing and split injection strategies on performance, emissions, and combustion characteristics of diesel engine fueled with biodiesel blended fuels. Fuel 213:106–114

Hurdogan E, Ozalp C, Kara O, Ozcanli M (2017) Experimental investigation on performance and emission characteristics of waste tire pyrolysis oil-diesel blends in a diesel engine. Int J Hydrogen Energy https://doi.org/10.1016/j.ijhydene.2016.12.126 (in Press Corrected Proof)

Jamrozik A (2017) The effect of the alcohol content in the fuel mixture on the performance and emissions of a direct injection diesel engine fueled with diesel- methanol and diesel-ethanol blends. Energy Convers Manage 148:461–476. http://dx.doi.org/10.1016/j.enconman.2017.06.030

Krishnamoorthi M, Malayalamurthi R (2017) Experimental investigation on performance, emission behavior and exergy analysis of a variable compression ratio engine fueled with diesel-aegle marmelos oil-diethyl ether blends. Energy 128:312–328. https://doi.org/10.1016/j.energy.2017.04.038

Kumar S, Dinesha P, Bran I (2017) Influence of nanoparticles on the performance and emission characteristics of a biodiesel fuelled engine: an experimental analysis. Energy 140(1):98–105

Mehregan M, Moghiman M (2018) Effects of nano-additives on pollutants emission and engine performance in a urea-SCR equipped diesel engine fueled with blended-biodiesel. Fuel 222:402–406

Musthafa MM, Kumar TA, Mohanraj T, Chandramouli R (2018) A comparative study on performance, combustion and emission characteristics of diesel engine fuelled by biodiesel blends with and without an additive. Fuel 225:343–348

Nair JN, Kaviti AK, Daram AK (2017) Analysis of performance and emission on compression ignition engine fuelled with blends of Neem biodiesel. Egypt J Pet 26(4):927–931

Patil AR, Desai AD, Madavi AD, Kamble SA, Navale SB, Dhutmal VU (2018) Comparative study on effect of biodiesel on CI engine performance and emission characteristics. Mater Today Proc 5(2):3556–3562

Perumal V, Ilangkumaran M (2017) Experimental analysis of engine performance, combustion and emission using pongamia biodiesel as fuel in CI engine. Energy 129:228–236. https://doi.org/10.1016/j.energy.2017.04.120

Perumal V, Ilangkumaran M (2018) Water emulsified hybrid pongamia biodiesel as a modified fuel for the experimental analysis of performance, combustion and emission characteristics of a direct injection diesel engine. Renew Energy 121:623–631

Puhan S, Jegan R, Balasubbramanian KG, Natarajan G (2009) Effect of injection pressure on performance, emission and combustion characteristics of high linolenic linseed oil methyl ester in a DI diesel engine. Renewable Energy 34:1227–1233. https://doi.org/10.1016/j.rene.2008.10.001

Rajak U, Verma TN (2018a) Effect of emission from ethylic biodiesel of edible and non-edible vegetable oil, animal fats, waste oil and alcohol in CI engine. Energy Convers Manag 166:704–718

Rajak U, Verma TN (2018b) Spirulina microalgae biodiesel—a novel renewable alternative energy source for compression ignition engine, J Cleaner Production 201:343–357. https://doi.org/10.1016/j.jclepro.2018.08.057

Ramesh DK, Kumar JLD, Kumar SGH, Namith V, Jambagi PB, Sharath S (2018) Study on effects of alumina nanoparticles as additives with poultry litter biodiesel on performance, combustion and emission characteristic of diesel engine. Mater Today Proc 5(1):1114–1120

Sander A, Košćak MA, Kosir D, Milosavljević N, Vuković JP, Magić L (2018) The influence of animal fat type and purification conditions on biodiesel quality. Renew Energy 118:752–760. https://doi.org/10.1016/j.renene.2017.11.068

Satputaley SS, Zodpe DB, Deshpande NV (2016) Performance, combustion and emission study on CI engine using microalgae oil and microalgae methyl esters. J Energy Inst 90:513–521. https://doi.org/10.1016/j.joei.2016.05.011

Singh TS, Verma TN, Nashine P, Shijagurumayum C (2018) BS-III Diesel Vehicles in Imphal, India: an emission perspective. In: Sharma N, Agarwal A, Eastwood P, Gupta T, Singh A (eds) Air pollution and control. Energy, environment, and sustainability. Springer, Singapore. ISBN: 978-981-10-7184-3 https://doi.org/10.1007/978-981-10-7185-0_5

Soto F et al (2018) The determination of the activation energy of diesel and biodiesel fuels and the analysis of engine performance and soot emissions. Fuel Process Technol 174:69–77

Suresh M, Jawahar CP, Richard A (2018) A review on biodiesel production, combustion, performance and emission characteristics of non-edible oils in variable compression ratio diesel engine using biodiesel and its blends. Renew Sustain Energy Rev 92:38–49

Tamilselvan P, Nallusamy N, Rajkumar S (2017) A comprehensive review on performance, combustion and emission characteristics of biodiesel fuelled diesel engines. Renew Sustain Energy Rev 79:1134–1159. http://dx.doi.org/10.1016/j.rser.2017.05.176

Vassilev SV, Vassileva CG (2016) Composition, properties and challenges of algae biomass for biofuel application: an overview. Fuel 181:1–33. https://doi.org/10.1016/j.fuel.2016.04.106

Rajak U, Verma, TN (2018) Effect of emission from ethylic biodiesel of edible and non-edible vegetable oil, animal fats, waste oil and alcohol in CI engine, Energy Conversion Manage 166:704–718. https://doi.org/10.1016/j.enconman.2018.04.070

Wei L, Cheung CS, Ning Z (2018) Effect of biodiesel-ethanol and biodiesel-butanol blends on the combustion, performance and emissions of a diesel engine. Energy 155:957–970

Wu Q, Xie X, Wang Y, Roskilly T (2018) Effect of carbon coated aluminum nanoparticles as additive to biodiesel-diesel blends on performance and emission characteristics of diesel engine. Appl Energy 221:597–604

Yao C, Cheung CS, Cheng C, Wang Y, Chan TL, Lee SC (2008) Effect of Diesel/methanol compound combustion on Diesel engine combustion and emissions. Energy Convers Manage 49:1696–1704

Yatish KV, Lalithamba HS, Suresh R, Hebbar HRH (2018) Optimization of bauhinia variegate biodiesel production and its performance, combustion and emission study on diesel engine. Renew. Energy 122:561–575

Zaharin MSM, Abdullah NR, Naja G, Sharudin H, Yusaf T (2017) Effects of physico- chemical properties of biodiesel fuel blends with alcohol on diesel engine performance and exhaust emissions: a review. Renew Sustain Energy Rev 79:475–493. https://doi.org/10.1016/j.rser.2017.05.035

Zareh P, Zare AA, Ghobadian B (2017) Comparative assessment of performance and emission characteristics of castor, coconut and waste cooking based biodiesel as fuel in a diesel engine. Energy 139:883–894. http://dx.doi.org/10.1016/j.energy.2017.08.040

Chapter 8
Review on the Use of Essential Oils in Compression Ignition Engines

S. M. Ashrafur Rahman, T. J. Rainey, Z. D. Ristovski, A. Dowell, M. A. Islam, M. N. Nabi and R. J. Brown

Abstract Essential oils are obtained from the non-fatty parts of a plant, such as the roots, bark, leaves, stems and flowers. These oils are mainly used in the natural medicine sector due to claimed health benefits, as well as the flavouring and fragrance sector, and the market has experienced rapid growth in recent years. The high quality required of the products leads to a very significant low-value waste stream, which is available for use in the transport and agricultural sectors. The use of essential oils in the compression ignition (CI) engine is a concept that has not yet been explored thoroughly. This paper analyses the available literature on the effect of essential oils and their blends on the performance, combustion characteristics and emission parameters of the CI engine. Regarding their properties, essential oils have similar properties to neat diesel. Engine performance using several essential oils and their blends improve brake thermal efficiency (BTE) and reduce brake-specific fuel consumption. A significant reduction in particulate matter (PM) emissions along with reduced hydrocarbon and carbon monoxide emissions has been reported in the literature. Furthermore, essential oils and their blends increase peak cylinder pressure and heat release rate (HRR) compared to neat diesel and biodiesel. However, most of the literature reported increased emissions of nitrogen oxides attributed to the lower cetane number and higher oxygen content. Due to having a low cetane number, essential oils can be used in CI engines by blending with either

S. M. Ashrafur Rahman (✉) · T. J. Rainey · Z. D. Ristovski · M. A. Islam · R. J. Brown
Biofuel Engine Research Facility, Queensland University of Technology (QUT), Brisbane, Australia
e-mail: smashrafur.rahman@hdr.qut.edu.au

Z. D. Ristovski
International Laboratory for Air Quality and Health, Queensland University of Technology (QUT), Brisbane, Australia

A. Dowell
Southern Cross Plant Science, Southern Cross University, PO Box 157 Lismore, NSW 2480, Australia

M. N. Nabi
Central Queensland University, Perth, WA, Australia

A. K. Agarwal et al. (eds.), *Methanol and the Alternate Fuel Economy*, Energy, Environment, and Sustainability,
https://doi.org/10.1007/978-981-13-3287-6_8

diesel or biodiesel. The rapid growth of the essential oil sector increases the likelihood of their utilization in CI engines in the future.

Keywords Essential oils · Diesel · Engine performance · Emission
Combustion · Fuel properties

8.1 Introduction

The progress of a nation critically depends upon its transportation, industry and power generation sectors. Diesel engines, also known as compression ignition (CI) engines, have become a source of power for these sectors due to having innate high efficiency, durability and better power output. Hence, they have a large share of the passenger car and heavy machinery market (Knecht 2008). With their rapid growth in demand, diesel engines are also becoming a significant source of pollutants. Both the environment and human health are adversely affected by the pollutants resulting from combustion of petroleum-derived fuel. For the past few decades, researchers are looking for ways to improve fuel properties to mitigate these problems. Some of the ways of modifying fuel properties to achieve improved combustion efficiency, and reducing emissions include blending biofuel with diesel and using fuel additives, amongst others. Biofuels are generally renewable and sustainable. There have been many studies that have focused on evaluating the performance of diesel engines using biofuel; mostly focused on biodiesel and bioethanol (Islam et al. 2017; Agarwal et al. 2011; Fattah et al. 2013, 2014; Palash et al. 2013). However, the introduction of essential oils as a diesel fuel substitute is an idea that has not been thoroughly exploited in the past. The production of essential oils is generally undertaken using decentralized agricultural methods. Harvesting and processing mostly use diesel engines. The waste stream derived from essential oil production is ideally located for use by farmers directly in their own diesel engines. The aim of this paper is to review the growing body of literature on the use of essential oils and their blends in diesel engines. At first, some elementary information about essential oils will be provided followed by in-depth analysis of the engine performance and emission of these oils in diesel engines.

8.2 Essential Oils

Essential oils are natural volatile renewable oils, responsible for many of the fragrances produced by plants. Vegetable oils (triglycerides) are usually derived from the fatty parts of a plant, such as the kernel, seed or nuts, whereas essential oils are obtained from non-fatty parts, for example, the roots, bark, leaves, stems and aromatic parts (flowers, petals and buds). Table 8.1 categorizes essential oils according to which part of the plant they are derived from.

Table 8.1 Parts of plants and corresponding essential oil examples

Parts of plant	Essential oil example
Leaves	Bay, Bay Laurel, Bergamot Mint, Cajeput, Cinnamon, Eucalyptus, Geranium, Hemp, Lemon Myrtle, Tea Tree, Violet
Flowers/petals/buds	Boronia, Cananga, Chamomile, Clove, Davana, Helichrysum, Hemp, Jasmine, Neroli, Rose
Wood	Amyris, Cedarwood, Palo Santo, Rosewood, Sandalwood
Bark	Cassia, Cinnamon
Grass	Citronella, Lemongrass, Palmarosa
Peel	Orange

In recent years, the essential oil market has grown rapidly. The global demand grew from 165 to 179 kilo-tonnes between 2014 and 2015, respectively (Grandview-Research 2015, 2016). This demand is predicted to surge at a compound annual growth rate of 8.4% from 2016 to 2024 (Grandview-Research 2016). This growth is accredited to the possible health benefits of essential oils, for example as complementary medicine, which has a history going back three thousand years. Some of the claimed benefits are shown in Table 8.2.

Table 8.2 Claimed health benefits of various essential oils (Steps 2017; Jeena et al. 2013; Stea et al. 2014; Zu et al. 2010; Lawless 2013)

Essential oil	Claimed health benefits
Basil	Mental alertness; supports sore muscles/joints, menstruation, etc.
Bergamot	Skin purifying, emotional support, etc.
Cedarwood	Support healthy skin
Clove	Antibacterial, anti-parasitic and antioxidant protection, local anaesthesia
Cypress	Soothes muscle tension
Eucalyptus	Improves respiratory issues such as bronchitis, sinusitis, and allergies. Also, invigorating and purifies the body
Ginger	Reduces inflammation, supports joints, improves digestion and relieves nausea
Lavender	Helps with relaxation, improves mood and heals burns and cuts
Lemon	Improves lymph drainage and cleanses the body
Lemongrass	Healthy digestion supports muscles and joints
Orange	Antioxidant, cleansing, purifying, etc.
Rosemary	Can naturally thicken hair, Improves brain function and memory
Rose	Reduces skin inflammation, creates glowing skin
Tea tree oil	Natural antibacterial, antifungal, reduces bad odours and can help stimulate the immune system
Sandalwood	Natural aphrodisiac that improves libido and can also improve energy

8.3 Chemical Composition of Essential Oils

Chemical components of essential oils are split into two categories: terpene-related hydrocarbons and highly oxygenated compounds. Terpenes are derived from isoprene. When terpenes are modified chemically through oxidation or reorganization of the carbon skeleton, they become terpenoids or isoprenoids. There are two types of terpenes usually found in essential oils, monoterpenes (10 carbons and 2 isoprene units) and sesquiterpenes (15 carbons and 3 isoprene units; shown in Fig. 8.1). Monoterpenes are a viable option for replacing gasoline, diesel and especially jet fuels, due to having a low freezing point and high energy density (due to methyl branching and cyclic structures) (Mendez-Perez et al. 2017). In contrast, Table 8.3 lists different oxygenated compounds present in various essential oils.

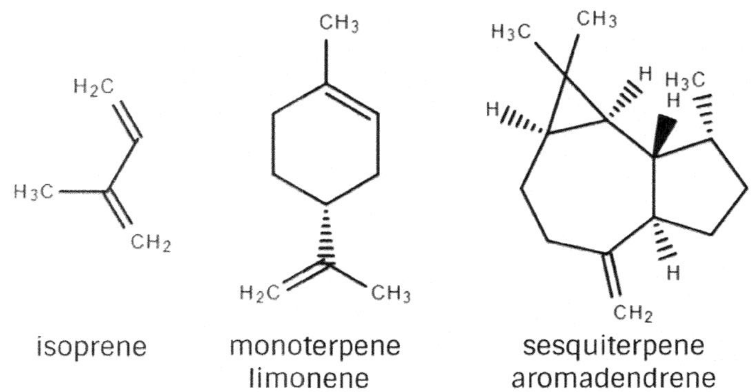

Fig. 8.1 Isoprene, mono- and sesquiterpene

Table 8.3 Different oxygenated compounds found in essential oils (Stewart 2005)

Oxygenated compound		Example	Essential oils
Esters		Linalyl acetate	Bergamot, Clary sage, Lavender
		Geraniol acetate	Sweet Marjoram
Aldehyde		Citral	Lemongrass, Melissa
		Citronellal	Melissa, Lemongrass, Lemon, Mandarin
Ketones		Thujone	Sage
Alcohols	Terpene	Linalool	Rosewood, Lavender
		Citronellol	Rose, Lemon, Eucalyptus
		Terpinen-4-ol	Tea tree
	Sesquiterpene	Bisabolol	Chamomile
Phenols		Eugenol	Clove, Cinnamon
		Thymol	Thyme
Oxides		Cineole	Eucalyptus, Rosemary, Cinnamon, Melissa

8.4 Essential Oil Extraction Process

The five main extraction processes that are used to obtain essential oils are water distillation, steam distillation, expression, solvent extraction and supercritical fluid extraction (carbon dioxide, CO_2).

8.4.1 *Water Distillation*

In this method, the plant material is immersed in water in an extraction chamber and then the still is heated. As the plant fibre is softened, the oil molecules are released as vapour, which rises with the steam. The oil vapours are carried to a condenser, which converts the steam to water and oil vapour to liquid essential oil. After that, the oil can be collected from a separator. This process is commonly known as hydro-distillation.

However, there are some disadvantages of this process:

- Oil components such as esters are sensitive to hydrolysis.
- Components such as acyclic monoterpene hydrocarbons and aldehydes are vulnerable to polymerization.
- Oxygenated components tend to dissolve in water.
- It is a very slow process and takes a long time to obtain enough oil.
- Large amounts of energy are required to heat the water.

8.4.2 *Steam Distillation*

This method is similar to the water distillation method, but, instead of water, steam is generated from an external supply. Higher pressure and higher temperature steam can be used for extracting. The advantage is that the steam supply (amount, temperature and pressure) can be controlled. and there is little thermal decomposition of oil components. However, this process has much higher capital expenditure. There is a variation of the steam distillation method, known as hydro-diffusion. In this method, the steam is fed through the top instead of the bottom of the still. Low pressures and low temperatures are extremely important in producing therapeutic grade oils.

8.4.3 Expression

The plant materials are mechanically pressed to express the oils. This process is used for plant materials that contain higher levels of essential oils, such as citrus fruit essential oils (e.g. orange, lime, lemon oils). An important consideration with pressed oils is the source of the crop since citrus fruits are constantly sprayed with pesticides, which can become very concentrated in the oil.

8.4.4 Solvent Extraction

A hydrocarbon solvent is used to dissolve essential oils along with plant material. After filtering the solution, it is concentrated by distillation. After distillation, a combination of wax and essential oil remains. Pure alcohol is then used to extract the oils or fats. When the alcohol evaporates, the residue is the essential oil. Solvent extraction is a method of extracting essential oils that are dominated by the perfume industry, which technically does not produce therapeutic grade oils, because in this process, solvents such as hexane, acetone and di-methylene-chloride are used. A major drawback of this process is the adverse health and environmental effects of the chemical solvents.

8.4.5 Supercritical CO_2 Extraction

This modern technique uses CO_2 to extract essential oils or resins from the plant when liquefied under pressure. Once the liquid depressurizes, the CO_2 returns to a gaseous state, and only pure essential oil, resins, or alkaloids remain. The CO_2 is a colourless and odourless liquid and can be completely removed; however, there is no research at this point to verify if there are any effects of using this procedure. Additionally, the extraction equipment is extremely expensive. So, the industry has not completely adopted this method and mostly uses steam distilled and expressed oils.

8.4.6 Summary of the Extraction Methods

Table 8.4 provides an overview of the extraction methods used for essential oils. It is apparent that steam distillation and expression are the most common methods. Blended (mixed) approaches are used for several essential oils.

Table 8.4 List of essential oils by extraction method

Extraction method	Essential oils
Steam distillation	Basil, Bay, Calamus, Camomile, Camphor, Caraway, Tea Tree, Thyme, Sandalwood, Rosewood, Roseberry, Rose, Pine, Nutmeg, Lemongrass, Lavender, Eucalyptus, Lime
Water distillation	Clove
Cold Pressing	Bergamot, Grapefruit, Lemon, Lime, Mandarin, Orange
Solvent	Benzoin
Mixed	
Solvent + distillation	Jasmine
Water + steam	Nutmeg

8.5 Performance of Essential Oils in a Diesel Engine

In this literature review section, nineteen research articles have been investigated to study the effects of essential oil and its blends on engine performance, combustion characteristics and emission parameters. This section is divided into two parts; the first part discusses the impact of essential oils when they are used in blends, and the second part discusses when essential oils are introduced as a diesel fuel additive. In order to define additives are substances that can be used in a very low percentage (<1%), which will vary engine performance and emission, whereas blends are used in a higher percentage (>1%). The first part is categorized according to cetane number of pure essential oils.

8.5.1 Essential Oils as Blends

Clove stem oil

Clove stem oil (CSO) is non-edible, and clove trees are abundant in East Africa and South Asia regions, for example, Tanzania (Mbarawa 2010). Clove stems are less commercial and can be used to produce oil, which is less expensive than pure clove oil (Mbarawa 2008). Major components of CSO are eugenol (81.49%) and caryophyllene (10.54%), while there are some minor components such as α-caryophyllene (2.16%), eugenol acetate (3.45%) and caryophyllene-oxide (2.37%). CSO has a very low cetane number, significantly lower calorific value and higher density compared to regular (base) diesel. For 25% CSO-diesel blends, the cetane number is only 35 compared to a cetane number of 51 for regular diesel (as shown in Table 8.5). Compared to diesel, CSO has a significantly higher flashpoint, which means the oil is safe to store and transport at room temperature.

The lower cetane number of CSO prohibits the use of pure oil and so can only be used by blending with diesel (Mbarawa 2008, 2010). As the CSO content in blended fuel increases, brake thermal efficiency (BTE) increases. This increase may

Table 8.5 Various physical and chemical properties of essential oils

Properties	Clove stem (Mbarawa 2010, 2008)	Tea tree (Rahman et al. 2017)	Pine (Vallinayagam et al. 2013a, b)	Orange (Rahman et al. 2017; Purushothaman and Nagarajan 2009; Govindarajulu et al. 2016; Deep et al. 2015; Rahman et al. 2017)	Turpentine (Anandavelu et al. 2010; Anand et al. 2010; Karthikeyan and Mahalakshmi 2007)	Mineral Turpentine (Karikalan and Chandrasekaran 2017)	Lemongrass (Sathiyamoorthi and Sankaranarayanan 2014; Alagumalai 2015; Sathiyamoorthi and Sankaranarayanan 2016, 2017; Ranitha et al. 2014)	Eucalyptus (Rahman et al. 2017; Vallinayagam et al. 2013; Devan and Mahalakshmi 2008; Tamilvendhan and Ilangovan 2011; Senthil et al. 2017; Verma et al. 2016)	Diesel standard		
									Aus (Australian Government 2017)	EU (EN590) (Last accessed 19 Sep 2017)	US (ASTM D957-17) (Standards 2017)
Density (kg/m^3)	1034.1 @20 °C	900 @15 °C	875.1 @15 °C	816.9 @30 °C 849.0 @15 °C	860.0–900.0 920	790.0 @15 °C	984.0 @27 °C 910.0	900.0, 913.0, 930.0	820–850	820–845	820–850
Viscosity (mm^2/s)	4.10 @20 °C	1.84 @40 °C	1.30	3.50, 0.96 @40 °C 3.52	2.50 @30 °C 3.89 @40 °C	3.88 @40 °C	4.18, 4.10 @40 °C	1.60, 2.00 3.2 @40 °C	2.00–4.50 @40 °C	2.00–4.50 @40 °C	1.90–4.10 @40 °C
Surface tension (m N/m)	32.1	29.4	–	27.7	–	–	–	28.5	–	–	–
Flashpoint (°C)	104.0^a, 114.0^c	53.5^c	52.0^h	55.0^c, 60.5^c, 61.0^e, 74.0^i	38.0^i	36.0^i	50.0^i	39.0^i, 53.0^i, 54.0^h, 55.0^c, 58.0^i, 93.0^i, 98.0^i	61.5 (min)	55.0 (min)	52.0 (min)
Pour point (°C)	−57	–	–	–	–	–	–	−12	–	–	–
Fire point (°C)	–	–	–	58, 82	–	–	58	–	–	–	–
Auto ignition temp (°C)	–	–	–	300	325	–	–	300–330 425	–	–	–

(continued)

Table 8.5 (continued)

Properties	Clove stem (Mbarawa 2010, 2008)	Tea tree (Rahman et al. 2017)	Pine (Vallinayagam et al. 2013a, b)	Orange (Rahman et al. 2017; Purushothaman and Nagarajan 2009; Govindarajulu et al. 2016; Deep et al. 2015; Rahman et al. 2017)	Turpentine (Anandavelu et al. 2010; Anand et al. 2010; Karthikeyan and Mahalakshmi 2007)	Mineral Turpentine (Karikalan and Chandrasekaran 2017)	Lemongrass (Sathiyamoorthi and Sankaranarayanan 2014; Alagumalai 2015; Sathiyamoorthi and Sankaranarayanan 2016, 2017; Ranitha et al. 2014)	Eucalyptus (Rahman et al. 2017; Vallinayagam et al. 2013; Devan and Mahalakshmi 2008; Tamilvendhan and Ilangovan 2011; Senthil et al. 2017; Verma et al. 2016)	Diesel standard		
									Aus (Australian Government 2017)	EU (EN590) (Last accessed 19 Sep 2017)	US (ASTM D957-17) (Standards 2017)
Boiling point (°C)	–	–	150–180	176	150–180	–	–	156, 175, 178	–	–	–
Cetane number	Negligible[f]	10.4[b]	11.0[d, g]	19.5[b], 47.0[i, f]	38.0[a]	–	38.0[g, i], 45.0[i]	Less than 15[d, g] 10.0[b], 18.0[i], 28.0[d], 53.0[i], 55.0[i]	51.0 (min)	51.0 (min)	40 (min)
Heating value (MJ/kg)	33.60	41.90	42.80	34.65, 40.38, 41.67	37.28, 44.40, 45.06	46.00	36.28, 36.00	43.27[j], 44.10	–	–	–
Lubricity (micron)	315	–	–	–	–	–	–	–	460 (max)	460 (max)	520 (max)
Carbon content (% Weight)	73.2	82.9	–	84.3, 87.4	88.2	–	79.9	77.9, 79.9	–	–	–
Hydrogen content (% weight)	7.3	11.7	–	11.7, 12.5	11.8	–	10.6	11.6, 11.7	–	–	–
Oxygen	19.5	5.4	–	0.1, 3.1	–	–	9.5	8.4, 10.4	–	–	–
C/H ratio	10.0	7.1	–	6.8	–	–	7.5	6.7	–	–	–

[a]ASTM (not specified); [b]ASTM D710; [c]ASTM D93; [d]ASTM D976; [e]IS 15607 Standard; [f]ASTM D613; [g]Cetane Index; [h]ASTM D92; [i]Unspecified; [j]Lower heating value

be associated with higher oxygen content, which in turn improves combustion efficiency and the energy output of CSO-diesel blends. However, fuel consumption and brake-specific fuel consumption (BSFC) increased and specific energy consumption decreased due to the lower energy heating value, and higher viscosity (Mbarawa 2008, 2010). Due to the lower heating value, more fuel was needed to produce a brake power similar to that produced by neat diesel (Mbarawa 2008). An increase of CSO concentration in the blend reduces fuel cetane number and results in a consequently longer ignition delay period. The researchers have speculated that it results in the air–fuel mixture becoming leaner than the lean combustion limit due to the excessive air, which then results in increased carbon monoxide (CO) emissions (Mbarawa 2010). In contrast, a reduction of CO is reported in another study, which was attributed to higher oxygen content in the blend, promoting oxidation of CO (Mbarawa 2008). Oxides of nitrogen (NO_X) emissions were found to increase with an increase in CSO content, which was attributed to higher oxygen content in the fuel blend and higher in-cylinder temperature (Dorado et al. 2003; Yu et al. 2002; Gomez 2000). Furthermore, fuels with ring structure result in higher adiabatic temperature (Delucchi 1997). However, hydrocarbon (HC) emissions decreased as there was more oxygen available for improved combustion when CSO content increased. The presence of oxygen improves diffusion combustion, which results in a significant reduction of smoke emission. The authors suggested that for long-term engine durability, tests are needed to ensure CSO does not adversely affect the engine and combustion and heat release rate (HRR) analysis of CSO-diesel blends. The authors concluded that though commercial CSO blending has a positive impact on engine performance and emission, it is not an economically viable option due to its high price.

Tea tree oil
Tea tree oil is produced from leaves of the *Melaleuca alternifolia*, which is native to Southeast Queensland and the Northeast coast of New South Wales, Australia (Wikipedia Contributors 2017). This oil has an oxygen content of around 5%. The main constituents of tea tree oil are terpinen-4-ol (41.66%), γ-terpinene (21.5%), and α-terpinene (9.69%) (Rahman et al. 2017). Compared to base diesel, tea tree oil has higher density, lower viscosity, slightly lower flashpoint, lower heating value, significantly lower cetane number and induction time (Rahman et al. 2017).

Compared to base diesel, tea tree oil–diesel blends emit higher CO emissions, which may be associated with higher density and lower heating value. Tea tree oil contains oxygen in OH form, thus, during combustion, the –OH radical reacts with nitrogen to form nitric oxide (NO) thus increases NO_X emission, compared to base diesel. Tea tree oil is less viscous and has a lower cetane number. This results in longer ignition delay; ignition moves more towards top dead centre (before TDC). Thus, combustion products have shorter residence time at elevated temperatures, which results in less oxidation of soot precursors (Vallinayagam et al. 2015). As a result, particulate matter (PM) emissions increase.

Pine oil

Pine oil is produced by a steam distillation method from stumps and was mainly used by Native Americans (Pine Essential Oil 2017). It mainly consists of α-terpineol or cyclic terpene alcohols. Pine oil has lower viscosity, cetane number and flashpoint and higher density compared to base diesel (Vallinayagam et al. 2013).

Increasing the pine oil content in blended fuels increases the peak HRR. This may be associated with a longer ignition delay due to the lower cetane number. This leads to accumulation of combustible mixture in the combustion chamber. As a result, a higher HRR is achieved. Also, lower bulk modulus and compressibility results in the delayed onset of combustion (Rakopoulos 2013). Furthermore, enhanced vaporization and improved combustion (due to having lower viscosity and boiling point) result in a higher peak combustion pressure for pine oil blends. This lower viscosity can contribute to finer atomization of fuel droplets, particularly in pre-common rail diesel engines, and hence improved combustion. Also, pine oil contains oxygen, which promotes combustion. Thus, BTE of pine oil blends were higher compared to neat diesel (Vallinayagam et al. 2013). Regarding exhaust gas temperature (EGT), increase in pine oil content in blends reduced EGT, which may be associated with better combustion, as supported by another study (Saravanan et al. 2009). Higher premixed HRR increases the combustion temperature, which results in increased NO_X emission for pine oil blends. However, the relation between increase in combustion temperature and reduction of EGT is not clear at this stage. It is reported that CO emission from pine oil is load dependent, at lower loads, the presence of oxygen dilutes air–fuel mixtures, affecting combustion and resulting in higher CO emission. At higher loads, improved combustion results from additional oxygen and reduced viscosity reduces the CO emission. Similar results were also reported in another study (Vallinayagam et al. 2013). In this study, NO_X emissions follow different trends at different loads. At higher loads, higher NO_X is emitted while at lower loads, lower NO_X was emitted by pine oil. Diesel produced higher NO_X at lower load as it has relatively higher fuel-to-air ratio, which predominates the combustion process at extreme low load conditions. The presence of oxygen in pine oil promotes combustion, which in turn results in lower smoke emission, as supported by another study (Yoshimoto et al. 2013).

Orange oil

Orange oil is extracted from *Citrus sinensis*, which is native to China, but is now cultivated extensively in Australia and America; the oil mostly contains d-limonene (>90%) (Esoteric Essential Oils 2017). From the literature, two types of orange oil have been tested in diesel engines. One type of orange oil contained 84.28% carbon, 12.47% hydrogen and 3.05% oxygen (Purushothaman and Nagarajan 2009) and another one contained negligible oxygen (Rahman et al. 2017). Most of the literature reports similar or higher heating value for orange oil compared to base diesel; however, some variants have significantly lower heating value too. Contrary results regarding cetane number are also found in the literature. Some authors claim a significantly lower cetane number (Rahman et al. 2017), and some claim it has a cetane number similar to that of diesel (Purushothaman and Nagarajan 2009).

Blending orange oil significantly reduced the viscosity and density and increased the heating value. However, it also resulted in a decrease in flashpoint and fire point temperature. Due to having lower viscosity and higher calorific value, blends with orange oil exhibit higher BTE (Purushothaman and Nagarajan 2009). Orange oil results in higher peak cylinder pressure, which can be credited to higher flame velocity ensuring better combustion (Poola et al. 1994). Furthermore, longer ignition delay for neat orange oil results in more fuel burning throughout the premixed combustion phase, consequently increasing the peak pressure and the maximum rate of pressure rise (Huang et al. 2004). Higher HRR for neat orange oil at the premixed combustion phase is attributed to the presence of oxygen in the oil. Neat orange oil also exhibits better evaporation, which along with longer ignition delay, increases the maximum HRR. In another study, higher BTE and lower brake-specific energy consumption (BSEC) for neat orange oil due to improved evaporation and mixing resulted in better combustion (Kumar et al. 2001).

Neat orange oil reduced CO emissions due to having lower viscosity and higher oxygen content, which results in complete combustion. The smaller fuel droplets mix thoroughly throughout the combustion chamber, which ensures improved combustion. Also, improved combustion results in lower HC and smoke emissions for neat orange oil. However, NO_X emissions are higher because of the higher oxygen content and higher cylinder pressure resulting in higher combustion temperature. A significant reduction in CO and NO_X emissions for orange oil blends was reported in a different study (Govindarajulu et al. 2016). However, poor injection characteristics and improper mixing of blended fuels with air during the first phase of combustion result in increased HC emissions for orange oil blended fuel. Furthermore, the blends increased BSFC (slightly), PM emissions and particle number.

Turpentine oil
Non-mineral Turpentine

Turpentine is a thin, volatile, essential oil, which is distilled from the resin of certain pine trees (Ryman 2002). These trees grow around the Mediterranean. Also, the kauri or *Agathis australis* from New Zealand produces turpentine essential oil. Compared to diesel, non-mineral turpentine oil has higher calorific value, higher density, lower cetane number and lower viscosity.

Turpentine oil blends reduced BSFC due to having better combustion (Anandavelu et al. 2010) and slightly increased BTE compared to neat diesel (Anandavelu et al. 2010; Anand et al. 2010). Increased BTE for turpentine oil is attributed to its lower viscosity and higher calorific value. However, higher BSFC was reported in the literature, which is due to the combined effects of the relative fuel density, viscosity and heating value of the fuel (Anand et al. 2010). The higher density of blends containing a higher percentage of turpentine has led to more discharge of fuel for the same displacement of the plunger in the fuel injection pump. These factors also contribute to higher specific energy consumption.

Turpentine blends reduce EGT, which may be associated with lower cetane number and better combustion. The efficient mixing of turpentine oil with air, forming lean localized regions, limits primary smoke formation, which is responsible for lower smoke emissions. Due to having better combustion, lower HC and CO emissions were reported. Regarding NO_X emissions, contrary results were found in the literature (Anandavelu et al. 2010; Anand et al. 2010). Due to increased combustion temperature, turpentine oil blends emitted more NO_X than neat diesel. However, low-temperature combustion will lower NO_X emissions. Heat released during the premixed combustion period is higher than that of diesel oil due to a shorter delay period. On the other hand, turpentine oil–diesel blends reduced the cylinder pressure.

In dual fuel mode, at higher load, turpentine oil–diesel blends increase BSFC and reduce BTE, which is due to knocking. Knocking causes poor fuel utilization and increased cylinder temperature, which results in increased CO emissions (Karthikeyan and Mahalakshmi 2007). Furthermore, dual fuel mode turpentine–diesel blends increase NO_X emissions, which may be associated with the higher amount of heat released during the second stage of combustion, higher combustion temperature and higher heat content of turpentine. However, these factors result in reduced smoke emissions in dual fuel mode.

Mineral Turpentine

Mineral turpentine is chemically very different from non-mineral turpentine oil (essential oil). Mineral turpentine is a mixture of aliphatic and alicyclic C7–C12 hydrocarbons with a maximum content of 25% of C7–C12 aromatic hydrocarbons and a maximum benzene content of 0.1% by volume, whereas non-mineral turpentine essential oil largely consists of pinene and has inferior solvent properties compared to mineral turpentine. Though mineral turpentine is derived from crude oil, this oil is still discussed in this review paper to evaluate the performance of different types of turpentine oils available.

Compared to diesel, mineral turpentine has lower viscosity and density and slightly higher heating value. However, the flashpoint is quite low, which means it is more volatile (Karikalan and Chandrasekaran 2017). Also, when compared to jatropha biodiesel, the same trend is found. Jatropha biodiesel has a significantly higher flashpoint compared to both base diesel and mineral turpentine. Mineral turpentine was used to improve the engine performance of jatropha biodiesel with respect to base diesel (Karikalan and Chandrasekaran 2017). Compared to pure jatropha biodiesel, as mineral turpentine concentration increased in the blend, BSFC is reduced and BTE is increased, which authors claimed to be due to improvement in the heating value. Mineral turpentine reduces CO, NO_X and smoke emission, however, increases HC emission. Reduction of NO_X is attributed to lower combustion temperature. The authors claimed that the reduction of CO emission at full load is due to two reasons: improved combustion due to easier atomization and mixing of vegetable oils and easier burning of vegetable oils at higher temperature, due to having higher oxygen content. Reduction of HC emissions was also reported, due to having a higher oxygen concentration. However, the authors did not report oxygen concentration of any of the fuel blends trialed.

Lemongrass oil

Lemongrass is a genus of Asian, African, Australian and tropical island plants in the grass family (Wikipedia Contributors 2017). In India, lemongrass is widely used in various medicines. The oil is extracted from fresh or partly dried leaves by steam distillation. It has comparable heating value to base diesel, which enables it to be used in diesel engine although it has a lower cetane number.

Compared to base diesel, lemongrass oil–diesel blends increased BTE and BSFC (Sathiyamoorthi and Sankaranarayanan 2014). The increase of BTE is attributed to improved combustion due to better vaporization and increased BSFC is attributed to the lower calorific value. A significant reduction in HC emission is reported in the literature, which may be associated with longer ignition delay and accumulation of fuel in the combustion chamber. However, lemongrass oil blends emitted higher NO_X compared to diesel fuel, which is due to higher peak flame temperature. In an experiment, a significant increase in CO emissions is reported. Lemongrass oil–diesel blends increased cylinder pressure, as these blends absorb more heat from the cylinder immediately after injection due to faster and complete combustion and thus resulted in higher cylinder pressure. However, in another study, it was reported that premixed neat lemongrass oil resulted in lower in-cylinder pressure and a lower rate of heat release for neat lemongrass oil compared to that of diesel, which is attributed to having higher viscosity and lower calorific value (Alagumalai 2015). Also, lower ignition delay and higher combustion duration were reported for neat lemongrass oil. The lower ignition delay is attributed to ignition of lighter compounds formed through thermal cracking of neat lemongrass oil at higher cylinder gas temperatures.

Eucalyptus oil

Eucalyptus trees are native to Australia and are cultivated worldwide (Wikipedia Contributors 2017). The oil is extracted by distillation from leaves. The major component of eucalyptus oil is 1,8-cineole; which has low chemical reactivity and stability. Carbon atoms linked to endo-ether are fully substituted in the absence of carbon–carbon double bonds (Devan and Mahalakshmi 2008). Compared to base diesel, eucalyptus oil has a higher self-ignition temperature, lower heat content and a much lower cetane number. In order to use eucalyptus oil in a diesel engine, either preheated air has to be added with pure oil (Tamilvendhan and Ilangovan 2011) or it must be blended with diesel (Devan and Mahalakshmi 2008). An interesting result reported in the literature is that neat eucalyptus oil has much lower oxidation stability, but when it is mixed with diesel, it significantly improves the oxidation stability of neat diesel (Rahman et al. 2017). Oxidation stability is a factor of major importance regarding diesel engine fuels, since it can drastically affect the quality of fuels and their performance in CI engines. When eucalyptus oil is blended with annona biodiesel, the addition of eucalyptus oil reduces viscosity and improves volatility (Senthil et al. 2017).

A reduction of BSFC and BSEC compared to base diesel is reported, which is caused by eucalyptus having higher calorific value and lower density leading to improved atomization (Devan and Mahalakshmi 2008; Senthil et al. 2017). Higher BTE for eucalyptus oil blends were reported, which is attributed to lower viscosity and higher calorific value (Devan and Mahalakshmi 2008). In another study, BTE for eucalyptus oil was found to be load dependent. BTE was lower at lower loads and higher at higher loads compared to base diesel. At higher loads, due to having higher heat content, higher volatility and lower viscosity, BTE of eucalyptus was slightly higher than for neat diesel (Tamilvendhan and Ilangovan 2011). The lower viscosity of eucalyptus oil results in proper mixing and vaporization, improved spray formation and air entrainment. Eucalyptus oil is highly volatile and less viscous and has higher oxygen content; it burns quickly and releases heat in a shorter duration resulting in higher combustion temperature, especially at higher loads, which also results in higher EGT. Also, higher ignition delay was reported for eucalyptus oil (Tamilvendhan and Ilangovan 2011).

Regarding NO_X, the higher oxygen content and a higher amount of fuel burnt in premixed combustion results in increased emissions for eucalyptus oil blends. The positive side is that this also causes higher peak cylinder pressure, lower smoke, HC and CO emissions. Due to lower cetane number at higher blend percentages, at standard injection timing, 20% eucalyptus oil blend can be the best alternative (Devan and Mahalakshmi 2008). In contrast, regarding NO_X, a reduction of emissions is reported in another study where eucalyptus oils were used (Rahman et al. 2017). Due to complex structural bonding, for eucalyptus oil, oxygen is not able to react with nitrogen to form NO and other factors dominate combustion (such as lower cetane number), and thus less NO_X is emitted. Similar to tea tree oil, eucalyptus oil is also less viscous and has a lower cetane number, which results in longer ignition delay, and so ignition moves towards TDC. Thus, combustion products have less resident time at elevated temperature, which results in less oxidation of soot precursors (Vallinayagam et al. 2015). Consequently, PM emission increases. Also, the presence of aromatic (branched, ring, multi-ring or polycyclic) structures may result in higher PM emissions (Pullen and Saeed 2014; Pulagala et al. 2009).

8.5.2 *Essential Oils as Bio-additives*

Essential oils can be selected as bio-additives, as they are typically oxygenated and are cyclic compounds, which improve combustion. Furthermore, these oils have 10–20 carbons, which decrease the strength of Van der Waals bonds in diesel fuel to further improve combustion. These oils are a natural product, so using them as a bio-additive will help decrease dependency on synthetic additives.

Clove and turpentine oils were selected as additives to diesel because during combustion they increase the reactivity of diesel (Kadarohman et al. 2008). The main components of clove oil are eugenol (70.54%) and cariofilen (21.54%). The main component of turpentine oil is α-pinen (34.82%). Increase in clove oil percentage in the blend slightly increases specific gravity and decreases viscosity. The addition of turpentine oil also reduces the viscosity of diesel. Oxygen present in clove oil helped reduce fuel consumption by increasing combustion reactivity. Oxygen in clove oil and the bulky structure of compounds in turpentine oil increased combustion reactivity and improved combustion, which resulted in reduced CO, HC and smoke emissions; however, NO_X emissions were higher.

In another study, clove oil, eugenol and eugenyl acetate were selected as biofuel additives due to having a tendency to increase combustion reactivity (Kadarohman et al. 2010). Eugenol is the main component of clove oil, and eugenyl acetate is produced from eugenol through esterification. Eugenyl acetate has a higher oxygen content compared to the other two. From the results, these additives reduced BSFC at higher speed due to increased chemical reactivity. Further, the oxygen present in the structure helped oxidize CO to CO_2, resulting in reduced CO emissions. Eugenyl acetate, due to having higher oxygen content, achieved minimum CO and HC emissions compared to other bio-additives used in the study. With regard to smoke emissions, oxygen in the additive reduced emissions by oxidizing particulates. However, this resulted in higher NO_X emission compared to neat diesel.

This study was followed by another study, which investigated the combustion characteristics using 0.2% clove oil, eugenol and eugenyl acetate as an additive in diesel fuel (Kadarohman et al. 2012). Use of clove oil as an additive resulted in the lowest ignition delay. Researchers have proposed that due to having terpene compounds which acted as mediator agent between the bio-additive and base fuel resulting a perfect solution resulting in better mixing with diesel and thus improved the combustion. However, no further studies have supported this conclusion. Furthermore, compared to diesel, the bio-additives have higher oxygen content, which results in an increase of combustion reactivity. This reduces ignition delay and increases HRR. However, in these studies, the authors did not report oxygen contents of any of the additives or fuel with additives.

In Table 8.5, the properties of various essential oils used in a diesel engine are shown. Table 8.6 lists the engine performance, emission parameters and combustion characteristics of essential oil and their blends in a diesel engine. These two tables are also categorized according to the cetane number of pure essential oils.

Several researchers used biodiesel produced from eucalyptus to operate a diesel engine (Verma et al. 2016; Tarabet et al. 2012), and another researcher used eucalyptus biodiesel as primary fuel in a dual fuel mode engine with natural gas (Tarabet et al. 2014). However, the concern of this review paper is to discuss the use of essential oils in a diesel engine and their effect on engine performance, emission and combustion characteristics. Thus, the results of the above-mentioned papers are not discussed here, as they are out of context.

8.6 Other Approaches to Improve Essential Oil Performance in Diesel Engine

Several researchers tried to improve the performance of essential oils in a diesel engine by adding different types of additives and using different emission-reduction techniques. This section will review literature that focused on improving the essential oil performance.

Ethanol was used to improve the performance of lemongrass oil (Sathiyamoorthi and Sankaranarayanan 2017). On average, compared to 25% lemongrass oil–diesel blends, ethanol-added blends reduce BSFC by around 8% and increased BTE by 12.8%. The addition of ethanol also resulted in higher EGT. Ethanol has around 35% oxygen, which results in better combustion. Due to having a lower cetane number, ethanol-added blends exhibited higher ignition delay. Thus, during combustion, these blends will have additional oxygen to burn, which results in rapid pressure rise and peak cylinder pressure. Compared to 25% lemongrass oil–diesel blends, ethanol-added blends exhibited 5.2% increased peak pressure and 26% increase in maximum HRR. The rise of HRR is attributed to the preparation of a larger portion of fuel for rapid burning due to the longer ignition delay. Additional oxygen from the ethanol helps oxidation of CO, which results in lower CO emissions for ethanol-added blends. However, this results in higher in-cylinder temperature, which is one of the predominant factors resulting in higher NO_X emissions. Increasing in ethanol percentage in blend increased HC emission, which may be associated with the higher formation of lean flame out region resulting from higher ignition delay. The addition of ethanol in lemongrass oil–diesel blends results in an increased rate of diffusion and turbulent mixing of the unburned/partial burnt fuel, which helps reduce smoke emissions.

n-butanol was used to improve the performance of orange peel oil (Deep et al. 2015). The authors reported lower BTE and higher HC and smoke emissions. These may be associated with lower heating value and lower cetane number of OPO-n-butanol blends. The low cetane number deteriorates ignition characteristics of the blends and promotes the quenching effect in the leaner mixture zone of the cylinder resulting in higher HC emissions (Karabektas and Hosoz 2009). However, the authors found a reduction of CO and NO_X emissions. The reduction of CO emission is attributed to higher volatility and better atomization characteristics.

In another study, antioxidant additives were used to reduce the emissions of lemongrass oil–diesel blends (Sathiyamoorthi and Sankaranarayanan 2016). Two additives were used: butylated hydroxyanisole (BHA) and butylated hydroxytoluene (BHT). These additives were added with 25% lemongrass oil–diesel blends at 500, 1000 and 1500 ppm. The addition of BHA additives reduces BSFC giving a maximum reduction of 1.66%, whereas BHT additives exhibit only 0.96% maximum reduction. A slight increase in BTE for additive added blends was reported, which is attributed to higher power output and lower BSFC. Due to having antioxidant in blends, it prevents the fuel conversion process, which results in reduced EGT for additive added blends. BHA and BHT exhibit a maximum

Table 8.6 Effect of essential oils on engine performance, emission parameters and combustion characteristics of diesel engine

Essential oil	Blend/additive (%)	Other fuels	Engine spec[c]	Engine performance				Emission parameter				Combustion characteristics			Ref.
				BTE (%)	BSFC (g/kWh)	BSEC (MJ/kWh)	EGT (°C)	HC (ppm)	CO (%)	NO_X (ppm)	Smoke	PP (bar)	ID (° CA)	Peak HRR (kJ/m^3.deg)	
Blend															
Clove Stem	25 and 50%	Diesel	–,4, 2499, 18.4, Water, D, –, –	↑	↑6.4%	–	Similar	↓18%	↑11.6	↑12.6%	–	–	↑	–	(Mbarawa 2010)
				↑	↓	–	–	↓	↓	↑	↓ (%)	–	–	–	(Mbarawa 2008)
Tea tree	5 and 10%	Diesel	160, 6, 5900, 17.3, –, D, –, –	–	–	–	–	–	↑	=/↑	↑ (PM)	–	–	–	Rahman et al. (2017)
Pine	25, 50, 75 and 100%	Diesel	5.2, 1, –, 17.5, Water, D, 23, 220	↑	↓			↓ (higher load) ↑ (lower load)		↓ (lower load) ↑ (higher load)	↓ (HSU)	↑			Vallinayagam et al. (2013)
	10, 20, 30, 40 and 50%	Diesel	5.2, 1, –, 17.5, Water, D, 23, 220	↑	↓	–	↓	↓ (higher load) ↑ (lower load)		↑	↓45.5% (HSU)	↑	↑	↑	Vallinayagam et al. (2013)
Orange	10%	Diesel	160, 6, 5900, 17.3, –, –, –, –		↑	–	–	–	↓	↓	↑ (PN & PM) [a]	–	–	–	Rahman et al. (2017)
	15%	Diesel Pongamia	–, 1, –, 16.5, Water, 25, 200	↑1.5%	↓10%	–	↓	↑6.5%	↓	↓ [b]	–	–	–	–	Govindarajulu et al. (2016)
	100%	Diesel	–, 1, 661 cc, –, Air, D, 23, 215	↑	↓	↑	–	↓ [a]	↓ [a]	↑ [a]	↓ (BSU)	↑9.9%	↑	↑ > 50% (J/degree)	Purushothaman and Nagarajan (2009)
	5 and 10%	Diesel	160, 6, 5900, 17.3, –, D, –, –	–	–	–	–	–	↓ ↑ (full load)	↓	↑ (PM)	–	–	–	Rahman et al. (2017)

(continued)

Table 8.6 (continued)

Essential oil	Blend/additive (%)	Other fuels	Engine spec[c]	Engine performance				Emission parameter				Combustion characteristics			Ref.
				BTE (%)	BSFC (g/kWh)	BSEC (MJ/kWh)	EGT (°C)	HC (ppm)	CO (%)	NO_X (ppm)	Smoke	PP (bar)	ID (° CA)	Peak HRR (kJ/m^3.deg)	
Non-mineral Turpentine	–	Diesel (DF)	4.4, 1, 661, 17.5, Water, –, –, –	Similar, small ↓ @full load	Similar, small ↑ @full load		↑	↑	↑	↑ @full load	↓ (BSN)	↑	↑	↑(J/degree)	Karthikeyan and Mahalakshmi (2007)
	20, 40, 60, 80 and 100%	Diesel	5.2, 1, –, 17.5, Water, D, 23, 220	↓	↓	–	–	–	–	↑	↓ (HSU)	similar	↓	↑	Anandavelu et al. (2010)
	30, 40,570 and 100%	Diesel	5.2, 1, –, 17.5, Water, D, 23, 220	↓	↑	↑	↓	↓	↓	↓	↓ (HSU) ↓ PM (g/h)	Slightly lower	↑	↓	Anand et al. (2010)
Mineral Turpentine	10, 20, 30, 40 and 50%	Jatropha Biodiesel Diesel	–, 1, –, –, Air, –, –, –	↑	↓	–	–	↑	↓	↓	↓ (FSN)	–	–	↑	Karikalan and Chandrasekaran (2017)
Lemongrass	25, 50, 75 and 100%	Diesel	–, 1, 553, 16.5, Water, D, 23, –	↑7.7%	↑9%	–	~Same	↓	↑ higher load	↑	–	↑	–	↑54%	Sathiyamoorthi and Sankaranarayanan (2014)
	5 and 10% premixed 100%	Diesel	–, 1, –, 17.5, Air, 4.5 (Port fuel), 3.5 (Port fuel)	↑	↑	–	–	–	–	–	–	–	–	–	Alagumalai (2015)
Eucalyptus	20, 30, 40 and 50%	Diesel	–, 1, 661, 17.5, Air, D, 23, 200	↑	–	↓	↑	↓[a]	↓	↑[a]	↓ (BSN)	↑	↑	↑(J/degree)	Devan and Mahalakshmi (2008)
	100%	Diesel	–, 1, –, –, Air, D, –, –	↑1%	–	–	–	Same or ↓	↓	↑	↓45% (BSN)	–	↑	–	Tamilvendhan and Ilangovan (2011)
	50, 60, 70, 80 and 90% with Annona biodiesel	Diesel Annona Biodiesel	–, 1, 661, 17.5, –, –, –, 200	↓	↓	–	↓	↓	↓	↓	↑ (HSU)	–	–	–	Senthil et al. (2017)
	5 and 10%	Diesel	160, 6, 5900, 17.3, –, D, –, –	–	–	–	–	–	↓	↓	↑ (PM)	–	–	–	Rahman et al. (2017)

(continued)

Table 8.6 (continued)

Essential oil	Blend/additive (%)	Other fuels	Engine spec[c]	Engine performance				Emission parameter				Combustion characteristics			Ref.
				BTE (%)	BSFC (g/kWh)	BSEC (MJ/kWh)	EGT (°C)	HC (ppm)	CO (%)	NO_X (ppm)	Smoke	PP (bar)	ID (° CA)	Peak HRR (kJ/m^3.deg)	
Additive															
Clove	0.2%	Eug EugAc Diesel	–,1, 661, –, –, –, –, –	–	–	–	–	↓	↓	↑	↓ (n/a)	–	–	–	Kadarohman et al. (2010)
			–, 1, 661, –, –, –, –, –	–	–	–	–	–	–	–	–	–	↑	↑(J)	Kadarohman et al. (2012)
Clove Turpentine	0.2% Clove 0.6% Turpentine	Diesel	–, 1, 661, –, –, –, –, –	–	↓	–	–	↓	↓	↑	↓(n/a)	–	–	–	Kadarohman et al. (2008)

[a]Refers to g/kWh; [b]Refers to %

BSFC Brake-Specific Fuel Consumption, *BSEC* Brake-Specific Energy Consumption, *BSN* Bosch Smoke Number, *BSU* Bosch Smoke Unit, *BTDC* Before Top Dead Centre, *BTE* Brake Thermal Efficiency, *CA* Crank Angle, *cc* Cubic Centimetre, *CO* Carbon Monoxide, *CR* compression Ratio, *DF* Dual Fuel mode, *DI* Direct Injection, *EGT* Exhaust Gas Temperature, *FSN* Filter Smoke Number, *g/kWh* gram per kilowatt-hour, *HC* Hydrocarbon, *HSU* Hartridge Smoke Unit, *HRR* Heat Release Rate, *ID* Ignition Delay, *IP* Injection Pressure, *IT* Injection Timing, *J* Joule, NO_X Oxides of Nitrogen, *PM* Particle Mass, *PN* Particle Number, *PP* Peak Pressure, *ppm* Parts Per Million, % percentage

Eug Eugenol, *EugAc* Eugenyl Acetate

n/a → unit not given

[c]Engine specification engine power (kW), number of cylinder, engine size (cc), compression ration, cooling system, direct/indirect injection, IT (° BTDC), IP (bar)

reduction of 7.1 and 5.3% EGT, respectively. Antioxidant added blends reduce peroxyl and hydrogen peroxide radical concentration, which results in the delayed conversion of CO, thus increases CO emission. Furthermore, due to the absence of oxidative-free radicals, HC and smoke emission are also increased for antioxidant added blends. However, this absence of free radicals prevents NO_X formation; thus for antioxidant added blends, reduction of NO_X emission is observed.

Emulsified lemongrass oil can be used to improve performance (Sathiyamoorthi et al. 2017). Use of water emulsion results in increased heat absorption by vaporized water. This decreases the local adiabatic flame temperature, which helps in reducing NO_X emissions. The authors also studied the effect of adding cerium oxide and diethyl ether with exhaust gas recirculation (EGR) on emission parameters of lemongrass oil–diesel blends. Nano-emulsified blends have a lower heating value, which results in higher BSFC. The effect of micro-explosion of the water droplets helps to break the larger oil droplets into many smaller oil droplets, accelerating fuel evaporation and the mixing with air, thereby resulting in a rapid combustion process and a higher BTE for nano-emulsified blends. Nanoparticles and diethyl ether (DEE) both have higher oxygen content, which also promotes combustion and increases BTE. Compared to lemongrass oil–diesel blends, emulsified additive added blends exhibit higher ignition delay and higher cylinder pressure. Emulsified blends emit higher CO emissions compared to lemongrass oil–diesel blends as pre-combustion temperature reduces with the addition of water. Cerium oxide addition reduces CO emissions, which may be associated with the presence of oxygen and shorter ignition delay. Furthermore, cerium oxide is an oxidation catalyst, which lowers carbon combustion activation temperature and increases HC oxidation. Along with it, micro-explosion of emulsified blends promotes HC oxidation. In the combustion chamber, cerium oxide nanoparticles absorb oxygen, which reduces NO_X formation. Emulsification also helps in reducing NO_X emissions by reducing the cylinder temperature. Improved combustion resulting from emulsification of water and addition of nanoparticles results in reduced smoke emission.

Catalytic converter (CC) and selective catalyst reduction (SCR) technologies (reducing agent urea) were used to improve emission characteristics of pine oil (Vallinayagam et al. 2013). A reduction of 32.4% NO_X emissions was reported. Ammonia produced from urea through hydrolysis, and thermolysis reacts with NO and NO_2 to from N_2 and H_2O and thus reduces NO_X. Furthermore, the oxidation catalysts reduce smoke emission by oxidizing the soot particles. Implementation of SCR + CC also reduces HC and CO emission by oxidizing unburnt HC and CO present in the exhaust.

8.7 Research Gap

From the discussion above, the following research gaps are identified:

- Essential oils obtained from different sources might have different chemical composition that may affect the engine performance and emission parameters differently. Thus, the detailed chemical composition of essential oils needs to be studied thoroughly before using in a diesel engine.
- All the chemical properties should be reported at standard temperature.
- Most researchers used emission instruments, which are not quite sophisticated enough to measure engine emissions, and calibration information of the instruments is missing
- In most of the cases, researchers measured only smoke emission, which does not fully represent PM emission. So, from these studies, it is not possible to clearly identify the impact of various essential oils on PM emission
- The researchers mostly stated that the chemical fuel properties mostly affect the engine performance and emission parameters, e.g. oxygen content, viscosity, heating value, etc. But in most cases, oxygen content was not reported. Full detailed investigation on these aspects needs to be carried out in future.
- Investigation of morphology and nanostructure of diesel PM for different essential oils need to be carried to develop relationship between fuel oxygen content, chemical structure and particle characteristics.
- Detailed cost analysis should be carried out.

8.8 Conclusions

The idea of using essential oils as fuels for diesel engines has recently been investigated by researchers. Essential oil components are mainly divided into two categories: terpenes and oxygenated compounds. Compared to neat diesel, essential oils mostly have similar properties such as heating value, density, viscosity. However, the cetane numbers of most of the essential oils is quite low. Overall, it was found that essential oils have varied effects on engine performance and emissions of a CI engine and thus their merits need to be considered on a case-by-case basis. Regardless, no significant degradation of engine performance and emission parameters was reported in the literature. Though essential oils are largely used in the medical and natural therapy sector as well as fragrance and flavouring industry, it can be concluded that blending essential oils with biodiesel/diesel can be used in a diesel engine. Furthermore, performance can be further improved through using various additives.

Acknowledgements The authors would like to thank QUT for providing a Ph.D. scholarship (Queensland University of Technology Postgraduate Research Award).

References

Agarwal AK, Gupta T, Kothari A (2011) Particulate emissions from biodiesel vs diesel fuelled compression ignition engine. Renew Sustain Energy Rev 15:3278–3300

Alagumalai A (2015) Combustion characteristics of lemongrass (*Cymbopogon flexuosus*) oil in a partial premixed charge compression ignition engine. Alexandria Eng J 54:405–413

Anand BP, Saravanan C, Srinivasan CA (2010) Performance and exhaust emission of turpentine oil powered direct injection diesel engine. Renew Energy 35:1179–1184

Anandavelu K, Alagumurthi N, Saravanan CG (2010) Performance and emission evaluation of low heat rejection direct injection diesel engine fueled by Diesel–Turpentine oil blends. ASME 2010 international mechanical engineering congress and exposition, pp 1581–1587

ASTM Standards (2017) D975-17: standard specification for diesel fuel oils. Last accessed 19 Sep 2017. https://www.astm.org/Standards/D975.htm

Australian Government: Department of Environment and Energy. Diesel fuel quality standard. Last accessed 19 Sep 2017. http://www.environment.gov.au/topics/environment-protection/fuel-quality/standards/diesel

Deep A, Kumar N, Kumar M, Singh A, Gupta D, Patel JS (2015) Performance and emission studies of diesel engine fuelled with orange peel oil and n-butanol alcohol blends. SAE technical paper, 2015-26-0049

Delucchi MA (1997) The annualized social cost of motor-vehicle use in the U. S. based on 1990–1991 data: summary of theory, data, methods, and results. In: Greene DL, Jones DW, Delucchi MA (eds) The full costs and benefits of transportation. Springer, Berlin, Heidelberg

Devan PK, Mahalakshmi NV (2008) An experimental investigation on performance and emission characteristics of eucalyptus oil-diesel blends in a D.I. diesel engine. SAE technical paper 2008-01-0757

Dorado MP, Ballesteros E, Arnal JM, Gómez J, López FJ (2003) Exhaust emissions from a diesel engine fueled with transesterified waste olive oil. Fuel 82:1311–1315

Esoteric essential oils. Orange essential oil (sweet) information. Last accessed 20 Sep 2017. http://essentialoils.co.za/essential-oils/orange.htm

Esoteric essential oils. Pine essential oil and pine needle essential oil information. Last accessed 20 Sep 2017. http://essentialoils.co.za/essential-oils/pine.htm

Fattah IMR, Masjuki HH, Liaquat AM, Ramli R, Kalam MA, Riazuddin VN (2013) Impact of various biodiesel fuels obtained from edible and non-edible oils on engine exhaust gas and noise emissions. Renew Sustain Energy Rev 18:552–567

Fattah IMR, Masjuki HH, Kalam MA, Hazrat MA, Masum BM, Imtenan S, Ashraful AM (2014) Effect of antioxidants on oxidation stability of biodiesel derived from vegetable and animal based feedstocks. Renew Sustain Energy Rev 30:356–370

Gomez MEG et al (2000) Emission and performance characteristics of a 2 litre Toyota diesel van operating on esterified waste cooking oil and mineral diesel fuel. In: Sokhi RS et al (eds) Urban air quality: measurement, modelling and management: proceedings of the second international conference on urban air quality: measurement, modelling and management held at the Computer Science School of the Technical University of Madrid 3–5 Mar 1999. 2000, Springer, Netherlands: Dordrecht, pp 13–20

Govindarajulu K, Rajasekhar Y, Gundabattini E, Rajagopal TKR (2016) Experimental investigation on engine performance and emission characteristics using Pongamia-Orange Oil blends. Int J Renew Energy Res 6:229–236

Grandview-Research (2015) Essential oil market analysis by product (Orange, Corn Mint, Eucalyptus, Citronella, Pepper Mint, Lemon, Clove Leaf, Lime, Spearmint), by application (Medical, Food & Beverage, Spa & Relaxation, Cleaning & Home) and segment forecasts To 2022

Grandview-Research (2016) Essential oil market analysis by product (Orange, Corn Mint, Eucalyptus, Citronella, Peppermint, Lemon, Clove Leaf, Lime, Spearmint), by application

(Medical, Food & Beverage, Spa & Relaxation, Cleaning & Home) and segment forecasts to 2024
Huang Z, Lu H, Jiang D, Zeng K, Liu B, Zhang J, Wang X (2004) Combustion behaviors of a compression-ignition engine fuelled with diesel/methanol blends under various fuel delivery advance angles. Biores Technol 95:331–341
Islam MA, Heimann K, Brown RJ (2017) Microalgae biodiesel: current status and future needs for engine performance and emissions. Renew Sustain Energy Rev 79:1160–1170
Jeena K, Liju VB, Kuttan R (2013) Antioxidant, anti-inflammatory and antinociceptive activities of essential oil from ginger. Indian J Physiol Pharmacol 57(1):51–62
Kadarohman A, Hernani Khoerunisa F, Astuti RM (2008) Potency of clove oil and turpentine oil as a diesel fuel bioadditives and their performance on one cylinder engine. In: Proceeding of the international seminar on chemistry 2008, pp. 721. Jatinangor, 30–31 Oct 2008
Kadarohman A, Hernani, Khoerunisa F, Astuti RM (2010) A potential study on clove oil, eugenol and eugenyl acetate as diesel fuel bio-additives and their performance on one cylinder engine. Transport 25:66–76
Kadarohman A, Rohman I, Kusrini R, Astuti RM (2012) Combustion characteristics of diesel fuel on one cylinder diesel engine using clove oil, eugenol, and eugenyl acetate as fuel bio-additives. Fuel 98:73–79
Karabektas M, Hosoz M (2009) Performance and emission characteristics of a diesel engine using isobutanol–diesel fuel blends. Renew Energy 34:1554–1559
Karikalan L, Chandrasekaran M (2017) Performance and pollutants analysis on diesel engine using blends of Jatropha biodiesel and mineral Turpentine as fuel. Int J Environ Sci Technol 14:323
Karthikeyan R, Mahalakshmi N (2007) Performance and emission characteristics of a turpentine–diesel dual fuel engine. Energy 32:1202–1209
Knecht W (2008) Diesel engine development in view of reduced emission standards. Energy 33 (2):264–271
Kumar MS, Ramesh A, Nagalingam B (2001) Complete vegetable oil fueled dual fuel compression ignition engine. SAE technical paper 2001-28-0067
Lawless J (2013) The encyclopedia of essential oils: the complete guide to the use of aromatic oils in aromatherapy, herbalism, health, and well being. Conari Press
Mbarawa M (2008) Performance, emission and economic assessment of clove stem oil–diesel blended fuels as alternative fuels for diesel engines. Renew Energy 33:871–882
Mbarawa M (2010) The effect of clove oil and diesel fuel blends on the engine performance and exhaust emissions of a compression-ignition engine. Biomass Bioenerg 34:1555–1561
Mendez-Perez D, Alonso-Gutierrez J, Hu Q, Molinas M, Baidoo EE, Wang G, Chan LJ, Adams PD, Petzold CJ, Keasling JD, Lee TS (2017) Production of jet fuel precursor monoterpenoids from engineered Escherichia coli. Biotechnol Bioeng. https://doi.org/10.1002/bit.26296
Palash SM, Kalam MA, Masjuki HH, Masum BM, Fattah IMR, Mofijur M (2013) Impacts of biodiesel combustion on NO_x emissions and their reduction approaches. Renew Sustain Energy Rev 23:473–490
Poola RB, Nagalingam B, Gopalakrishnan K (1994) Performance studies with biomass-derived high-octane fuel additives in a two-stroke spark-ignition engine. Biomass Bioenergy 6:369–379
Pulagala V, Vasanta R, Nittalla S (2009) Effects of key properties of biodiesel fuels on injection, combustion, and emission characteristics of a DI-CI engine. Injection systems for IC engines One Bridcage Walk, Westminster, London, IMechE
Pullen J, Saeed K (2014) Factors affecting biodiesel engine performance and exhaust emissions–part I: review. Energy 72:1–16
Purushothaman K, Nagarajan G (2009) Performance, emission and combustion characteristics of a compression ignition engine operating on neat orange oil. Renew Energy 34:242–245
Rahman SMA, Van TC, Hossain FM, Dowell A, Islam MA, Nabi MN, Marchese AJ, Tryner J, Rainey TJ, Ristovski ZD, Brown RJ (2017) Experimental investigation of fuel properties and emission characteristics of essential oil blends in a multi-cylinder CI engine. Submitted to fuel

Rahman SMA, Hossain FM, Van TC, Dowell A, Islam MA, Rainey TJ, Ristovsku ZD, Brown RJ (2017b) Comparative evaluation of the effect of sweet orange oil-diesel blend on performance and emissions of a multi-cylinder compression ignition engine. AIP Conf Proc 1851:020007

Rakopoulos D (2013) Combustion and emissions of cottonseed oil and its bio-diesel in blends with either n-butanol or diethyl ether in HSDI diesel engine. Fuel 105:603–613

Ranitha M, Nour AH, Sulaiman ZA, Nour AH (2014) A Comparative study of Lemongrass (*Cymbopogon citratus*) essential oil extracted by microwave-assisted hydrodistillation (MAHD) and conventional hydrodistillation (HD) method. Int J Chem Eng Appl 5:104

Ryman D (2002) Danièle Ryman's Aromatherapy Bible: an encyclopaedia of plants and essential oils and how they help you. Piatkus

Saravanan S, Nagarajan G, Rao GLN (2009) Feasibility analysis of crude rice bran oil methyl ester blend as a stationary and automotive diesel engine fuel. Energy Sustain Dev 13:52–55

Sathiyamoorthi R, Sankaranarayanan G (2014) Experimental investigation of performance, combustion and emission characteristics of neat lemongrass oil in DI diesel engine. Int J Curr Eng Technol 3:25–30

Sathiyamoorthi R, Sankaranarayanan G (2016) Effect of antioxidant additives on the performance and emission characteristics of a DICI engine using neat lemongrass oil–diesel blend. Fuel 174:89–96

Sathiyamoorthi R, Sankaranarayanan G (2017) The effects of using ethanol as additive on the combustion and emissions of a direct injection diesel engine fuelled with neat lemongrass oil-diesel fuel blend. Renew Energy 101:747–756

Sathiyamoorthi R, Sankaranarayanan G, Pitchandi K (2017) Combined effect of nanoemulsion and EGR on combustion and emission characteristics of neat lemongrass oil (LGO)-DEE-diesel blend fuelled diesel engine. Appl Therm Eng 112:1421–1432

Senthil R, Silambarasan R, Pranesh G (2017) Exhaust emissions reduction from diesel engine using combined Annona-Eucalyptus oil blends and antioxidant additive. Heat Mass Transf 53:1105–1112

Stea S, Beraudi A, De Pasquale D (2014) Essential oils for complementary treatment of surgical patients: state of the art. Evid-Based Complement Altern Med 2014, Article ID 726341, 6 pages

Stewart D (2005) The chemistry of essential oils made simple: god's love manifest in molecules. Care publications

Tamilvendhan D, Ilangovan V (2011) A performance, emission and combustion investigation on hot air assisted eucalyptus oil direct injected compression ignition engine. Mod Appl Sci 5:53

Tarabet L, Loubar K, Lounici MS, Hanchi S, Tazerout M (2012) Eucalyptus biodiesel as an alternative to diesel fuel: preparation and tests on DI diesel engine. BioMed Res Int 235485

Tarabet L, Loubar K, Lounici M, Khiari K, Belmrabet T, Tazerout M (2014) Experimental investigation of DI diesel engine operating with eucalyptus biodiesel/natural gas under dual fuel mode. Fuel 133:129–138

Steps SB An introductory guide to 1000 s of uses for essential oils. Last accessed 19/09/2017. http://www.sustainablebabysteps.com/uses-for-essential-oils.html

US EPA. Diesel fuel standards. Last accessed 19 Sep 2017. http://www.epa.gov/diesel-fuel-standards

Vallinayagam R, Vedharaj S, Yang WM, Saravanan CG, Lee PS, Chua KJE, Chou SK (2013a) Emission reduction from a diesel engine fueled by pine oil biofuel using SCR and catalytic converter. Atmos Environ 80:190–197

Vallinayagam R, Vedharaj S, Yang WM, Lee PS, Chua KJE, Chou SK (2013b) Combustion performance and emission characteristics study of pine oil in a diesel engine. Energy 57:344–351

Vallinayagam R, Vedharaj S, Yang W, Roberts WL, Dibble R (2015) Feasibility of using less viscous and lower cetane (LVLC) fuels in a diesel engine: a review. Renew Sustain Energy Rev 51:1166–1190

Verma P, Sharma MP, Dwivedi G (2016) Potential use of eucalyptus biodiesel in compressed ignition engine. Egypt J Pet 25:91–95

Wikipedia contributors "Tea tree oil," Wikipedia, The Free Encyclopedia. Last accessed 20 Sep 2017. https://en.wikipedia.org/w/index.php?title=Tea_tree_oil&oldid=799591597

Wikipedia Contributors. Cymbopogon. Wikipedia, The Free Encyclopedia. Last accessed 20 Sep 2017. https://en.wikipedia.org/w/index.php?title=Cymbopogon&oldid=796712083

Wikipedia Contributors. Eucalyptus Wikipedia, The Free Encyclopedia. Last accessed 20 Sep 2017. https://en.wikipedia.org/w/index.php?title=Eucalyptus&oldid=800723231

Yoshimoto Y, Kinoshita E, Shanbu L, Ohmura T (2013) Influence of 1-butanol addition on diesel combustion with palm oil methyl ester/gas oil blends. Energy 61:44–51

Yu C, Bari S, Ameen A (2002) A comparison of combustion characteristics of waste cooking oil with diesel as fuel in a direct injection diesel engine. Proc Inst Mech Eng Part D: J Automobile Eng 216:237–243

Zu Y, Yu H, Liang L, Fu Y, Efferth T, Liu X, Wu N (2010) Activities of ten essential oils towards propionibacterium acnes and PC-3, A-549 and MCF-7 cancer cells. Molecules 15(5):3200–3210

Part IV
Utilization Aspects

Chapter 9
Laser-Ignited Engine Development for Adaptation to Hydrogen-Enriched Compressed Natural Gas (HCNG)

Rajesh Kumar Prasad and Avinash Kumar Agarwal

Abstract Among alternative fuels, natural gas is readily available in large quantities in acceptable composition; therefore, no further processing is needed such as in case of biofuels. Compressed natural gas (CNG) is the cheapest alternative fuel with quickest commercial implementation potential due to only slight modifications required in the existing engine hardware. Lean-burn concept is an effective approach to enhance engine's thermal efficiency and reduce exhaust emissions, but engine's combustion performance deteriorates because of slower flame speed and higher cyclic fluctuations. Increasing turbulence and using multiple spark plugs are conventional approaches to increase the flame speed but at the expense of volumetric efficiency and higher electrode erosion due to increased energy deposition under these conditions. Therefore, enriching natural gas with hydrogen is an effective way to increase the flame speed, while reducing the ignition energy requirements. Hydrogen enrichment of CNG reduces the volumetric heat content of the fuel, which results in relatively lower energy density of the test fuel. Therefore, for the same power output, complete combustion of HCNG reduces HC and CO emissions. Using laser ignition instead of conventional spark ignition system and use of HCNG, lean combustion limit could be extended further. Extended lean limit operation results in further reduction in emissions.

Keywords Conventional spark ignition · Laser ignition · CNG
Hydrogen · HCNG · COV_{IMEP} · Emissions

R. K. Prasad · A. K. Agarwal (✉)
Engine Research Laboratory, Department of Mechanical Engineering,
Indian Institute of Technology Kanpur, Kanpur 208016, India
e-mail: akag@iitk.ac.in

A. K. Agarwal et al. (eds.), *Methanol and the Alternate Fuel Economy*,
Energy, Environment, and Sustainability,
https://doi.org/10.1007/978-981-13-3287-6_9

9.1 Introduction

Transport policies in the world have been mainly driven by the need to reduce the harmful automotive emissions, particularly in the urban areas. A complex combination of several factors such as fluctuating prices of petroleum products, undesirable pollutant species, and resultant ecological imbalances has shifted the research focus on to new/renewable/alternative fuels. Thus, GHG emission reduction and security of energy supply are the two main drivers for the development of alternative fuels worldwide. In an optimized development scenario, possible contributions of biofuels, compressed natural gas (CNG), and hydrogen toward reduction in environmental complications have been identified. CNG has emerged as the most promising alternative fuel for transport sector because it is the cleanest hydrocarbon fuel, which possesses highest H/C ratio among all hydrocarbon fuels, high research octane number (RON), good anti-knocking properties, more abundant availability than petroleum, and it produces significantly lower regulated emissions (e.g., CO, CO_2) and non-methane hydrocarbons (NMHCs) (Helmreich and Lochner 2011). In addition to these advantages, there are some disadvantages such as slower flame velocity, large cycle-to-cycle variations (CCVs), poor lean-burn capability, and higher brake-specific fuel consumption (BSFC). To enhance the lean-burn capability and to increase the flame velocity, higher turbulence in the combustion chamber is essential; however, this measure results in higher heat losses from the cylinder walls due to higher convective heat transfer (Sorensen 2011). Hydrogen enrichment of CNG delivers higher laminar flame burning velocities since the flame speeds are 2.9 m/s for hydrogen and 0.38 m/s for CNG for stoichiometric mixtures. Thus, hydrogen-enriched CNG (HCNG) smoothens the cyclic fluctuations caused by the relatively poor lean-burn capability of CNG engines (Fayaz et al. 2012). HCNG mixtures can reduce emissions (CO, CO_2, CH_4, and NO_x) and improve the fuel economy and the thermal efficiency.

9.1.1 Current Energy Scenario

Modern living standards are putting an additional demand for energy. Global energy demand for transport sector is projected to increase by ~30% between 2014 and 2040. Today diesel accounts for ~35% and natural gas accounts for ~2% of the total energy used in the transportation sector worldwide, but this share will increase to ~40% for diesel and ~5% for natural gas by 2040.

In the present scenario, ~95% of current transportation energy needs are met by petroleum products. However, in foreseeable future, gasoline demand will flatten out as shown in Fig. 9.1a because fuel economy of vehicles is expected to improve rather rapidly. The demand for diesel will grow since the truck and marine business

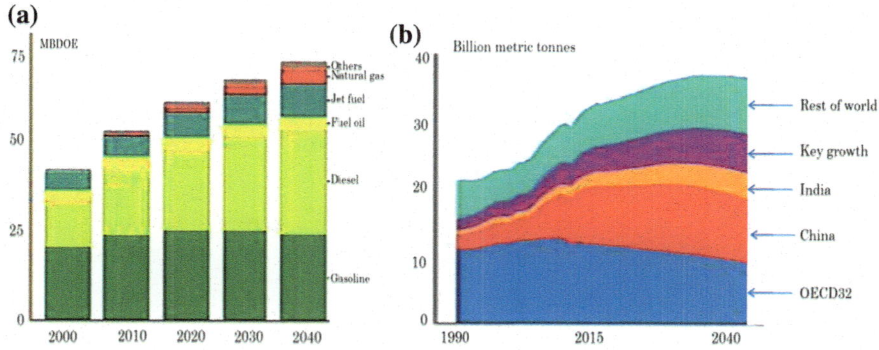

Fig. 9.1 **a** Global transportation fuel demand and **b** global CO_2 emissions (http://cdn.exxonmobil.com/~/media/global/files/outlook-for-energy/2016/2016outlook-for-energy.pdf)

needs to expand; hence, the natural gas consumption will grow mainly on account of the commercial fleets. Greenhouse gases (GHGs) emitted from the energy sector constitute two-thirds of total anthropogenic GHG emissions. In future, global CO_2 emissions will reduce (Fig. 9.1b) due to the use of different renewable energy sources such as solar, hydro, wind, nuclear along with the usage of carbon capture and storage techniques (Vander Hoeven 2015).

9.1.2 Natural Gas Outlook

The global market for natural gas is much smaller compared to that for petroleum-based fuels because its transportation economy is difficult. Only ~16% of global natural gas produced is internationally traded, and liquefied natural gas (LNG) accounts for less than 4%. Major gas reserves are located in the former Soviet Union, Canada, and the Middle East.

The OECD countries comprising mainly of developed countries (USA, Canada, Japan, South Korea, and others) and non-OECD countries (Russia, China, India, and others) consumed approximately 104 trillion cubic feet (TCF) natural gas during 2006. The annual global demand for natural gas is likely to rise to 153 TCF by 2030 (Fig. 9.2). Developing countries such as India and China are expected to considerably increase their natural gas consumption to match their expected domestic, transportation, and industrial growth. Presently, India and China utilize natural gas in small percentage of their total energy consumption matrix and are likely to show an annual increase of ~5% in natural gas consumption. Among non-OECD countries, the NG consumption is likely to grow by ~2.3% annually up to 2030, while the OECD countries will show relatively lower consumption growth rate of ~1.0% during the same period.

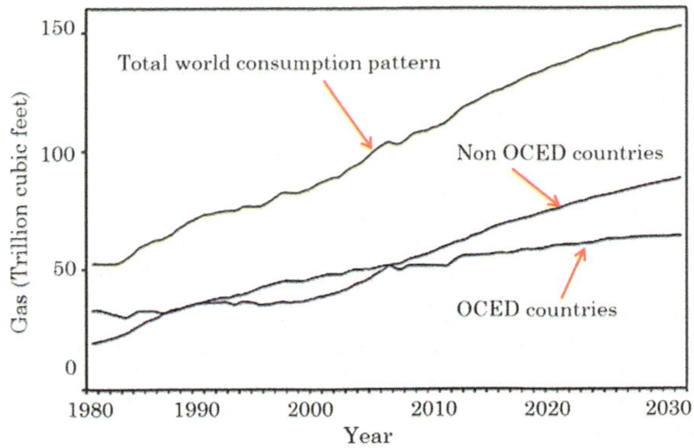

Fig. 9.2 Global natural gas (TCF) consumption (*Source* Energy Information Administration (EIA-2008)) (EIA U 2006)

9.1.3 Typical Fuel Properties

The H/C ratios in HCNG mixtures can vary from 4 to ∞ (v/v). Hythane is a specific HCNG mixture having 15% volume fraction H_2 in balance CNG, which was patented in 1992 by Frank Lynch of HCI (Karim 2003). Karner et al. (2003) investigated effect of H_2 enrichment of CNG on an AVL engine fueled by CNG, H_2, and different HCNG mixtures (20HCNG and 50HCNG). Since then, many engine researchers exhaustively investigated the impact of hydrogen enrichment of CNG on various engine combustion, performance, and emission characteristics.

Hydrogen fraction in HCNG mixture is controlled using Dalton's law of partial pressure (Ma et al. 2008). The physical, chemical, and other properties of HCNG mixtures fall in-between those of hydrogen and CNG, which contains ~95% methane (CH_4). Comparison of fuel properties of various SI engine fuels, namely hydrogen, methane (CNG), and gasoline, is given in Table 9.1 (Karim 2003).

9.1.4 HCNG Fuel Induction Technologies

In a conventional SI engine, homogeneous mixture of gasoline and air is inducted into the combustion chamber, compressed, and then ignited by a spark plug just before the expansion stroke begins. Figure 9.3 illustrates the basic landscape of HCNG engine technologies.

Gaseous fuel–air mixture formation outside the engine combustion chamber can be done by a separate mixing system or a port fuel injection (PFI) system. In the PFI system, HCNG is inducted upstream of the intake valve using a solenoid controlled gas injector, where fuel–air mixing occurs just before the entry into the engine cylinder during the intake stroke. A major drawback of this system is that fuel displaces the air during the intake stroke, resulting in lower volumetric and part load efficiencies. On the other hand, however, there are other challenges faced in development and use of DI injectors for hydrogen and HCNG (Kofler 2007) such as:

Table 9.1 Typical properties of SI engine fuels (Karim 2003)

Properties	Hydrogen	Methane	Gasoline
Flammability limits (v/v)	4–75	5.3–15	1.2–6
Stoichiometric composition in air (% v/v)	29.53	9.48	1.65
Stoichiometric fuel–air ratio (w/w)	0.029	0.058	0.0664
Minimum ignition energy (mJ)	0.02	0.28	0.25
Laminar flame speed at NTP (m/s)	1.90	0.38	0.37–0.43
Flame temperature (K)	2318	2190	2470
Auto-ignition temperature (K)	858	813	500–750
Density (gm/m^3)	82	717	5110
Heat of combustion (LHV), (MJ/kg)	119.7	46.72	44.79
Heat of combustion (HHV), (MJ/kg)	141.7	52.68	48.29
Specific heat ratio of NTP gas	1.383	1.308	1.05
Diffusion coefficient in air at NTP (cm^2/s)	0.61	0.189	0.05
Normalized flame emissivity	1.00	1.7	1.7

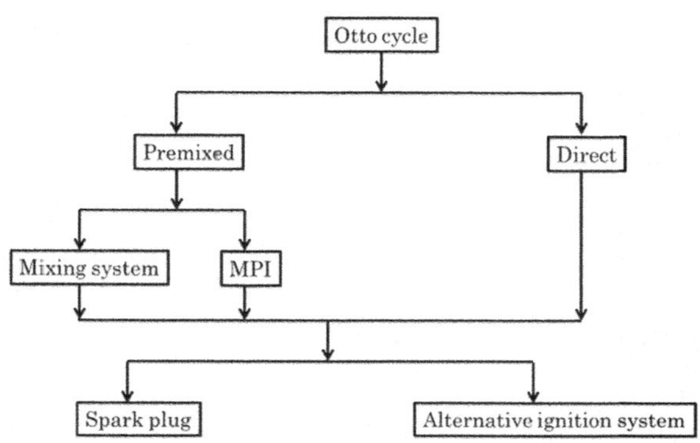

Fig. 9.3 Basic landscape of HCNG engine technologies (Raizer 1997)

- Reduced internal damping effect due to extremely low density of hydrogen.
- Reduced hydrodynamic lift due to very low viscosity of hydrogen.
- Hydrogen potentially degrades epoxy resins used to coat the piezo-stack.

Among the biggest challenges faced while investigating HCNG as a test fuel for IC engines is the design and development of a dynamic mixing system, fuel metering system and fuel injection system, which are essential to determine the most optimal H_2/CNG ratio, fuel quantity to be injected per minute, and the fuel injection pulse width. When the hydrogen fraction is increased above a certain limit, abnormal combustion such as preignition, knocking, and backfire may occur. For safe operation, both the spark timing and the fuel–air ratio are appropriately optimized. In addition to these, appropriate safety devices have to be installed in the fuel line in order to prevent any accident. As the hydrogen fraction in the HCNG mixture increases, lean limit extends and the maximum brake torque

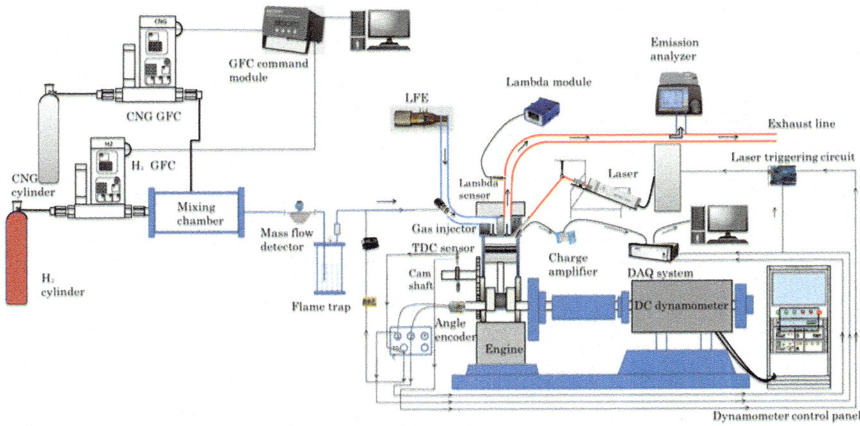

Fig. 9.4 Typical laser-ignited HCNG engine experimental setup (Prasad 2018)

(MBT) decreases. This indicated that there is a relationship between the hydrogen enrichment of CNG mixture, ignition timing, and excess air ratio. Lastly, the current cost of hydrogen is higher than that of CNG, rendering HCNG slightly more expensive than CNG (Karim 2003).

In a study, a naturally aspirated single-cylinder, water-cooled, compression ignition engine (Kirloskar, DM-10; 948 CC; CR: 11) was modified to operate on HCNG using laser ignition (Fig. 9.4). The engine was coupled to a DC dynamometer. This modified engine operated in both conventional spark ignition and laser ignition mode. A custom-built high-flow rate solenoid gas injector was used to supply HCNG into the intake port. An injector triggering circuit was designed using a microcontroller (Arduino; Uno) to actuate the injector, which generated a 5V TTL pulse that activated the injector driver module.

9.2 Ignition Systems

9.2.1 *Spark Ignition System*

In a conventional SI system, plasma is formed between the two electrodes, which ionize the combustible charge present between these electrodes. During the flame growth, these two electrodes restrict the growth in axial direction. Therefore, spark plug erosion is an undesirable consequence of interaction of electrical discharge plasma (high pressure and temperature) with the electrode surface. There are various mechanisms, by which an electrode material can erode, namely particle injection, vaporization of electrode material, and sputtering. The erosion rate of electrode depends on several factors such as specifications of the spark plug, gap between the electrodes, electrode material, electrode size and shape, combustion temperature, type of fuel used, and the manner in which energy is delivered by the spark plug. The reasons for spark plug erosion can be summarized as follows:

- Disintegration of the surface material under atomic or ionic bombardment (Bauer 1998).
- High current density in the arc phase heats up a small area of cathode material to its boiling point, leading to evaporation of some of the cathode material (Maly and Vogel 1979).
- Oxidation, when atoms in the electrode combine with oxygen, and a layer of metal oxides is formed on the surface of the electrodes (Jones 1963).

Necessary voltage required to produce a spark between the electrodes mainly depends on the pressure and temperature of the combustible charge at the time of ignition. As per Paschen's law, breakdown voltage primarily depends on the electrode gap and the cylinder pressure at the time of spark.

9.2.2 Laser Ignition System

Laser ignition is an innovative optical ignition concept with potential application in IC engines. The basic principle of laser ignition is rather simple, and not new. The following are the main advantages of laser ignition over a conventional spark ignition, which is typically used in all IC engines:

- Ignition of lean mixtures is feasible, which results in lower in-cylinder temperatures, leading to ultra-low NO_x emissions.
- No spark plug erosion effect, leading to significantly longer operational life span of laser ignition system.
- Choice of arbitrary plasma location into the engine cylinder.
- Multi-point ignition is possible, which speeds up the combustion for obtaining the highest engine efficiencies, especially for lean mixtures using a single laser source.
- Precise ignition timings are possible for obtaining optimal engine performance and maximum efficiency.
- Shorter ignition delay periods can be achieved.

Dale et al. (1978) performed the first laser ignition experiment in a gasoline engine as early as in 1978. They employed a pulsed CO_2 laser with laser pulse duration of 300 ns at wavelength of 10.6 μm. The laser beam was focused into the engine cylinder using 150-mm focal length lens to ignite both leaded gasoline

(RON 92) and unleaded gasoline. With the laser-ignited engine, continuous run time of ½ to 1¼ h was possible before soot deposits on the window surface led to misfiring in this study. For all λ tested, a more rapid pressure rise was obtained using 300 mJ pulse energy compared to a conventional spark plug. The conventional ignition system would not be able to operate the engine above $\lambda = 1.55$, while using laser ignition, the engine could be operated up to a $\lambda = 1.92$. Comparing the two ignition systems with 16% EGR, laser ignition showed ~17% improvement in BSFC. The laser-ignited engine produced higher NO_x emissions than conventional spark-ignited engine for $\lambda > 1.1$, which was not unexpected because of the superior engine performance of the laser ignition.

Taira (2015) compared conventional spark ignition system with microchip laser system (Fig. 9.5) in a production grade automotive engine (1AZ-FSE; Toyota), which was a 2.0-L, inline, four-cylinder GDI engine, operated using 1064-nm wavelength microchip laser, which generated 6 MW peak power at 500 ps. It was observed that laser ignition took much less energy (~2 mJ) to ignite the gasoline–air mixtures, compared to ~35 mJ energy required for the conventional spark plug system, and the flame speed was also faster in case of laser ignition system.

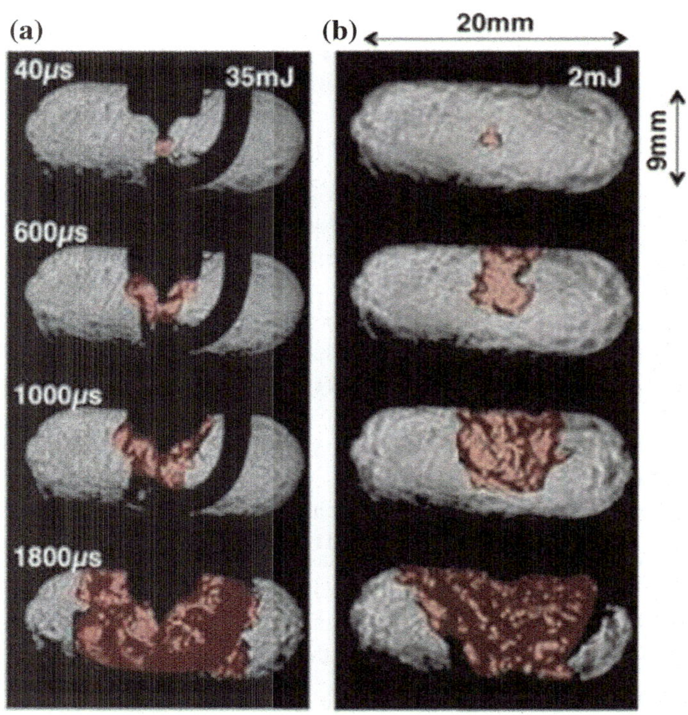

Fig. 9.5 Schlieren imaging of ignition by **a** a conventional spark plug and **b** a microchip laser in a customized test engine (Taira 2015)

9.2.3 Challenges of Laser Ignition

Non-availability of commercial compact and robust laser systems which could deliver the shortest possible laser pulses, at appropriate wavelength, high spatial and temporal coherence, and sufficient minimum pulse energy (MPE) for reliable ignition in all permissible modes of engine operation remains the foremost challenge. Laser ignition system must be highly resistant against forces, vibrations, and thermal loading, since these systems would be mounted on the cylinder head or the laser would be transported through the fiber-optical cable to the engine cylinder. Due to proximity of the engine cylinder's hot environment, appropriate cooling arrangement is needed for high-power laser, which might become a serious challenge for long run during summer. In addition, there are several other issues which are encountered and they are discussed briefly in the following paragraphs.

Laser beam transport: The method of laser beam transport becomes more complicated in multi-cylinder engines. Yalin et al. (2006) studied application of different types of fiber optics used for laser transport. They demonstrated propagation of laser beam through the fibers and a multiplexer system used for igniting multi-cylinder engine. It was reported that step-index silica fibers were inferior compared to coated hollow fibers for transport for laser pulses for engine applications. If optical fibers were to be considered for laser beam transport, it becomes rather critical that such high-intensity laser pulses do not destroy the structure of the fibers since energy intensities of the order of several GW/cm^2 are required to be transported (Phuoc 2000).

Influence of ambient temperature on solid-state Q-switched Nd:YAG laser performance: The laser spark plug works at temperatures up to 100 °C in engine-like conditions; hence, the laser performance can be affected by the ambient temperature in the engine proximity (Wintner and Kofler 2015). Wintner and Kofler (2015) measured energy content in different open and monolithic setups and reported that rise in ambient temperature lowers the energy content of the Q-switched pulses. Both the beam diameter and the mode order were major factors for reduced laser efficiency (Fig. 9.6). In this figure, RT(1) represents the condition of a wide and energetic laser beam, which operates in a perfectly cylindrical TEM_{10} mode.

In RT(4), the beam cross section reduced but it still operated in the same mode as observed at 100 °C (Wintner and Kofler 2015). Further increase in ambient temperature resulted in reduction in beam diameter as well as the pulse energy. It was reported that the energy intensity was the highest in the center, and it remained constant for all ambient temperatures. Additionally, laser pulse duration increased from 1.3 to 2.0 ns, if ambient temperature increased from room temperature to 200 °C. The numbers between the graphs indicate discrete points of measurement of beam profiles (@ 50 cm distance), from which the values of beam diameter were taken (monolithic resonator).

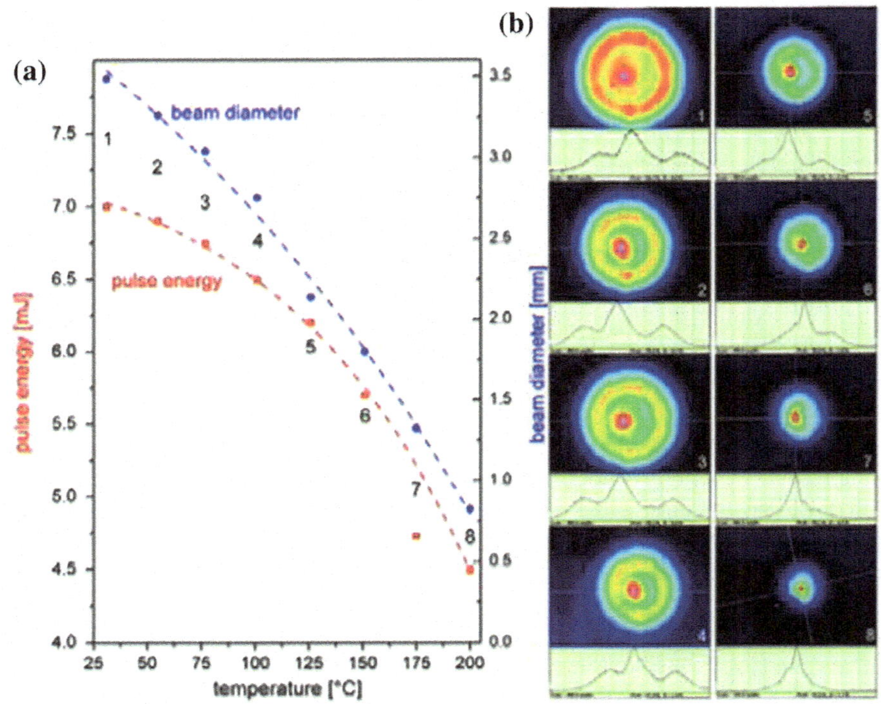

Fig. 9.6 Ambient temperature effect on **a** laser pulse energy and **b** laser beam profile (Wintner and Kofler 2015)

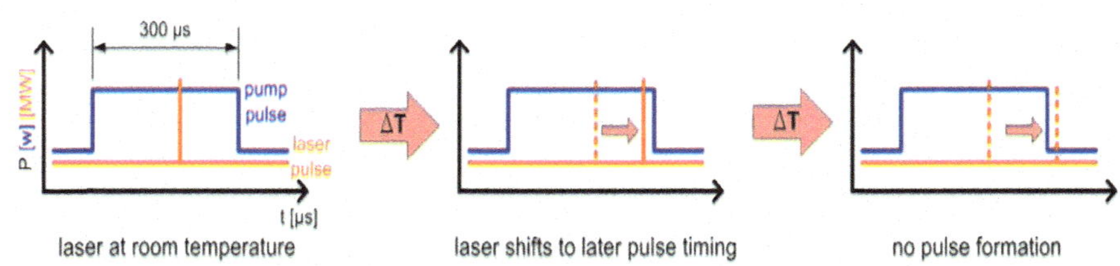

Fig. 9.7 Temperature effect on pulse formation timing at fixed geometry of laser crystal (Wintner and Kofler 2015)

Temperature dependence might critically change the laser ignition system's performance, thereby affecting its reliability for engine applications. With increasing ambient temperature, pulse forming shifts to later times due to more optical pumping energy stored in the resonator. Further increase in ambient temperature stops the pulse formation as shown in Fig. 9.7.

9.3 HCNG Engine Combustion, Performance, and Emissions

The effect of hydrogen enrichment of CNG affects the combustion, performance, and emission parameters and has been investigated for engine applications by several researchers in last few decades. The engine's fuel combustion efficiency is a direct indicator of degree of complete burning of HCNG mixtures in the engine combustion chamber (McNeil 2005).

9.3.1 HCNG Engine Combustion/Performance

Figure 9.8 shows the effect of hydrogen enrichment of CNG on the in-cylinder pressure variations at varying engine load conditions. Maximum in-cylinder

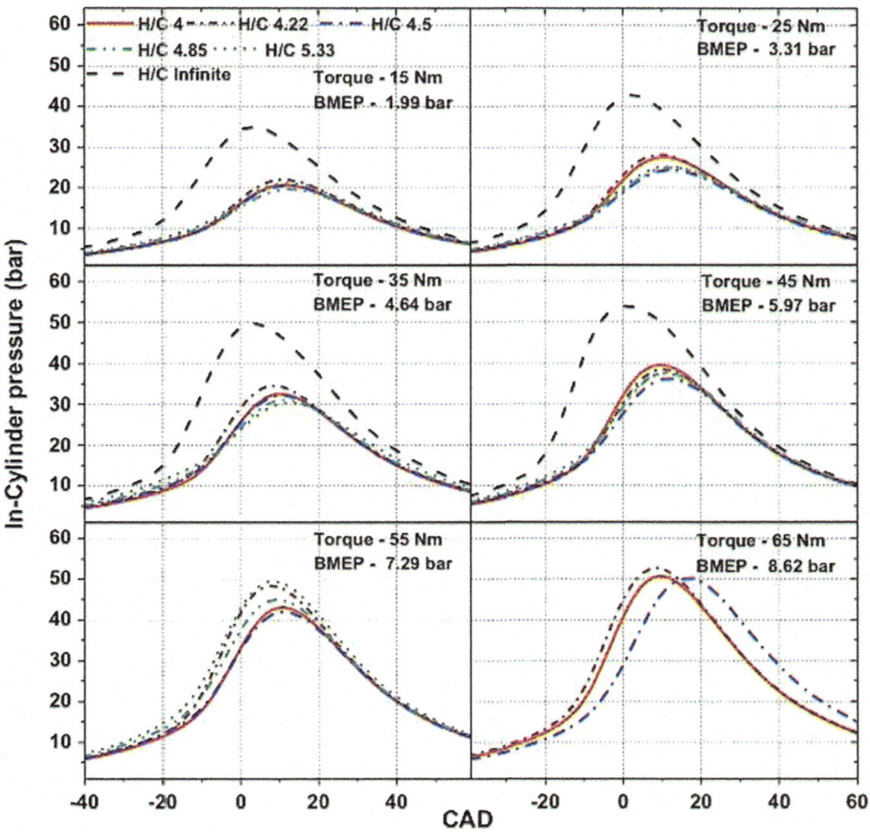

Fig. 9.8 Effect of hydrogen enrichment of CNG on the in-cylinder pressure variations (Verma et al. 2017)

pressure (P_{max}) for all HCNG mixtures increases with increasing BMEP. A noticeably higher combustion peak pressure was observed for hydrogen compared to all HCNG mixtures. This sharp increase in the combustion peak pressure for hydrogen was due to relatively early start of combustion (SoC) and higher flame speed. The peak cylinder pressure position (CAP_{max}) was almost constant at all loads. The engine torque output increased with increasing peak cylinder pressure, and the corresponding crank angle position remained almost constant for a fixed λ. For a fixed λ, the engine torque increased with hydrogen enrichment of CNG and was the lowest for hydrogen (45 Nm), which increased to 65 Nm for an H/C ratio of 4.5. H_2 enrichment of CNG was effective in increasing the engine torque up to H/C ratio of 4.5, and beyond this, there was sudden drop of 10 Nm in the engine torque.

Figure 9.9b shows the variations in engine combustion parameters such as SoC, EoC, CA_{50}, and combustion duration w.r.t. BMEP. SoC in case of hydrogen was $\sim 10°$ CA advanced compared to other HCNG mixtures. SoC and combustion duration (CD) were relatively longer at higher engine loads for all HCNG mixtures at MBT timing of 32° bTDC. SoC for hydrogen varied between 15° CA and 19° CA bTDC for the entire load range; however, for other test fuels, it varied from 1° CA aTDC to 9° CA bTDC. Combustion phasing in case of hydrogen advanced with increasing engine load, and it was significantly advanced compared to other HCNG mixtures. Karim et al. (1996) performed a theoretical study to evaluate the effect of hydrogen enrichment of CNG on engine's combustion characteristics at different spark timings and reported that both the ignition delay (ID) and the combustion duration (CD) decreased with hydrogen enrichment at identical λ. Raman et al. (1994) performed experimental investigations using HCNG mixtures with varying hydrogen fraction ranging from 0 to 30% (v/v) in a four-stroke Chevrolet Lumina, six-cylinder, water-cooled, V8, 3.135-L SI engine having a compression ratio of 8.8. Reduction in NO_x with slight increase in HC emissions for ultra-lean combustion of 15–20% (v/v) hydrogen-enriched CNG was reported. 15HCNG mixture showed relatively lower BSFC compared to CNG.

Figure 9.10 shows the effect of hydrogen enrichment of CNG and excess air ratio (λ) on the engine power output. Ma et al. (2010) reported that hydrogen enrichment of CNG resulted in relatively lower engine power output for $\lambda \leq 1.2$, as shown in Fig. 9.10, and it is because of higher heating value of CNG–air mixtures compared to HCNG–air mixtures. Upon increasing λ from 1.0 to 1.6, engine power output continuously drops for both CNG and HCNG mixtures. However, HCNG mixture with higher hydrogen fraction had the ability to increase the engine power output for leaner mixture. Furthermore, the power output from HCNG-fueled engines was significantly higher than CNG-fueled engines for $\lambda > 1.6$. Bysveen (2007) also reported that hydrogen enrichment of CNG increased the power output at lean limit operations. For 30HCNG, $\sim 30\%$ power was reduced when λ was increased from 1.0 to 1.8. Supercharging or turbocharging and use of higher compression ratio (CR) are generally used to extract more power output under extremely lean charge conditions. With increasing CR, heat release rate

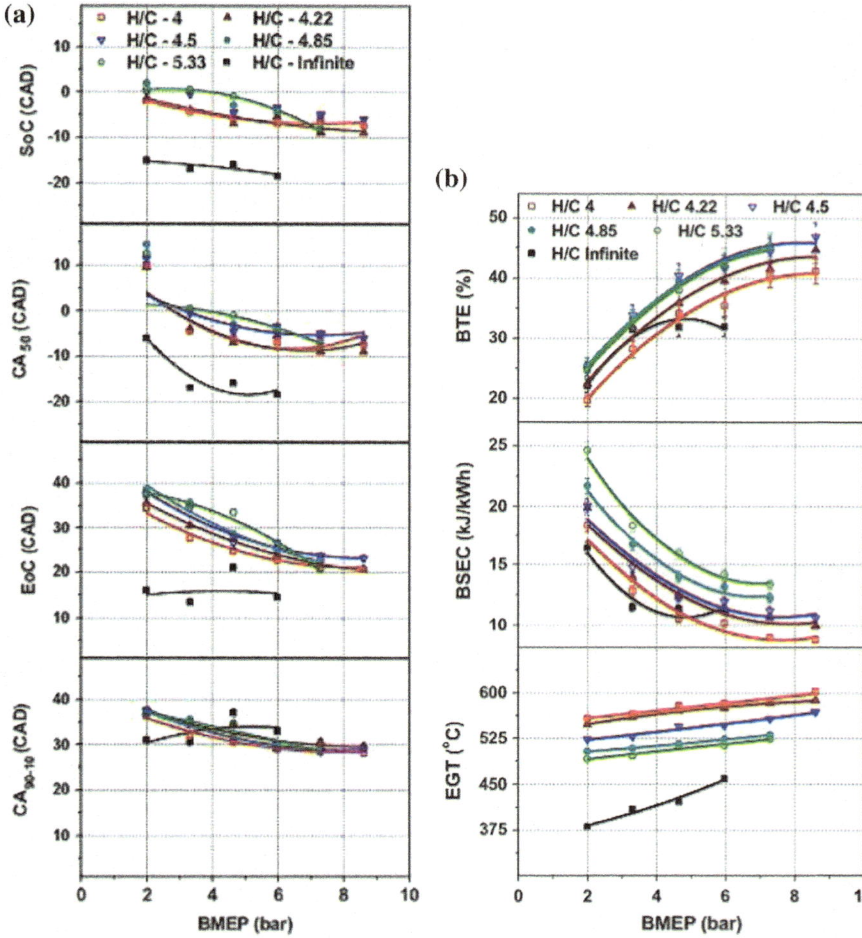

Fig. 9.9 **a** Variations of SoC, 50% MBF, EoC, and CD for HCNG mixtures, **b** variations in combustion parameters such as BTE, BSEC, and EGT for HCNG mixtures (Verma et al. 2017)

increases, leading to higher in-cylinder pressure and temperature, which results in increased thermal efficiency and power output.

Park et al. (2012) explored the effect of CR and CNG/30HCNG on torque variations with varying λ (Fig. 9.11) reported a significant reduction of engine torque for lean 30HCNG ($\lambda > 1.6$), when CR increased from 10.5 to 11.5. Figure 9.11 thus depicts the difficulty associated with lean-burn combustion, when the engine is operated at high CR for enhancing the performance. To overcome this difficulty, retarded spark ignition timing strategy was adopted (Park et al. 2012) to increase the power output from HCNG-fueled engine; however, this resulted in minor reduction in the thermal efficiency.

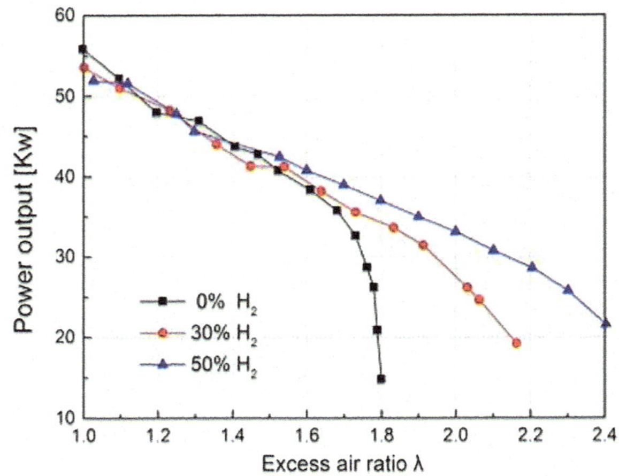

Fig. 9.10 HCNG engine power output w.r.t. λ and hydrogen fraction in the HCNG mixture (Ma et al. 2010)

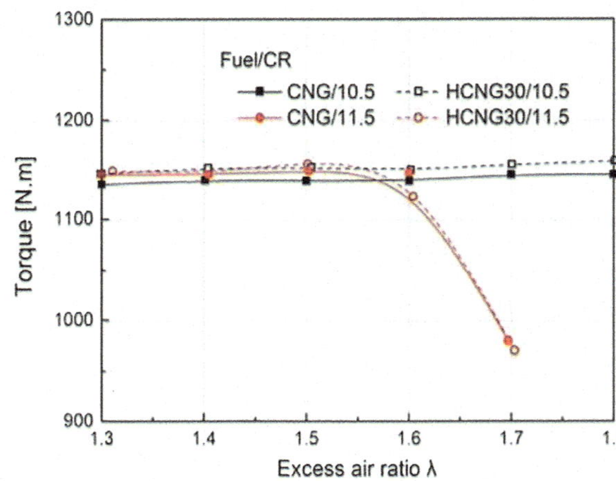

Fig. 9.11 HCNG engine torque for varying λ and compression ratio (Park et al. 2012)

Figure 9.12 depicts the effect of hydrogen enrichment of CNG on the indicated thermal efficiency (ITE) with varying λ. Ma et al. (2007) reported the effect of hydrogen enrichment on the ITE of a six-cylinder HCNG-fueled SI engine under varying λ conditions, with identical spark timing (Fig. 9.12).

It could be noticed that $\sim$50% hydrogen enrichment of CNG actually reduced the fuel economy compared to baseline CNG, when operated at $\lambda < 1.5$. Hydrogen enrichment reduced the engine's CD (increased the amount of heat addition), which is advantageous for the thermal efficiency; however, on the other hand, it reduced the quenching distance, thus increasing the convective heat transfer. Increased heat loss had a direct negative impact on the thermal efficiency (De Simio et al. 2013). With higher hydrogen enrichment of CNG, heat loss became dominant, resulting in a reduction in the thermal efficiency. Several other investigations have concluded that the increase in CR up to a certain limit can improve BSFC and power output (Ma et al. 2012; Lim et al. 2013; Tangöz et al. 2015).

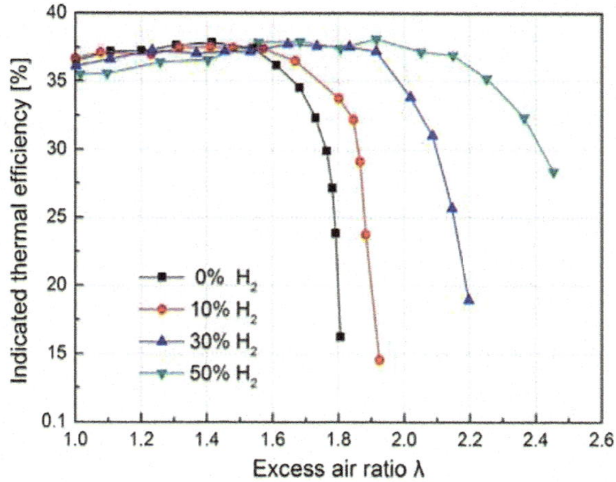

Fig. 9.12 Indicated thermal efficiency variations with excess air ratio (λ) and hydrogen enrichment of the CNG (Ma et al. 2007)

Lim et al. (2013) investigated the effect of compression ratio and observed increased ITE for both 30HCNG and CNG, when the CR was increased from 10.5 to 11.5 (Fig. 9.13). For the same CR, thermal efficiency was relatively higher for 30HCNG compared to CNG; however, the difference was rather small under lean condition (less than 1% for $\lambda < 1.5$). Tangöz et al. (2015) investigated the effect of CR on a 3.9-L engine operating at 1500 rpm and full load. BSFC initially reduced when CR was increased from 9.6 to 12.5, and then, it increased when CR was further increased from 12.5 to 15. Shudo et al. (2000) investigated the combustion and emission characteristics of DI stratified charge premixed HCNG–air mixtures under lean conditions in a CNG engine. It was reported that the engine delivered higher thermal efficiency due to hydrogen enrichment of CNG, which increased the flame speed and reduced the emissions such as HC and CO.

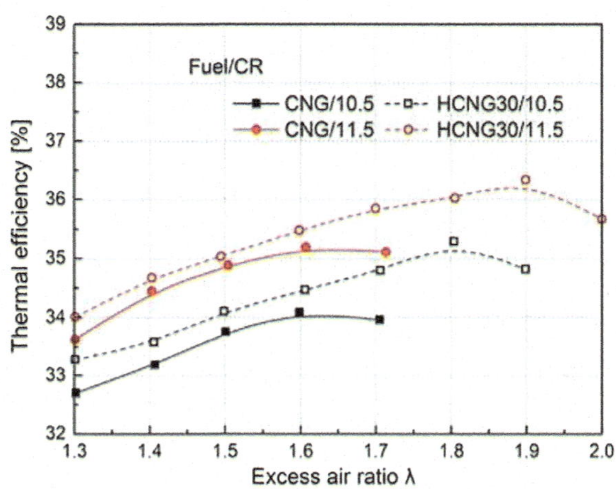

Fig. 9.13 Effect of CR on the thermal efficiency of CNG/30HCNG with varying λ (Lim et al. 2013)

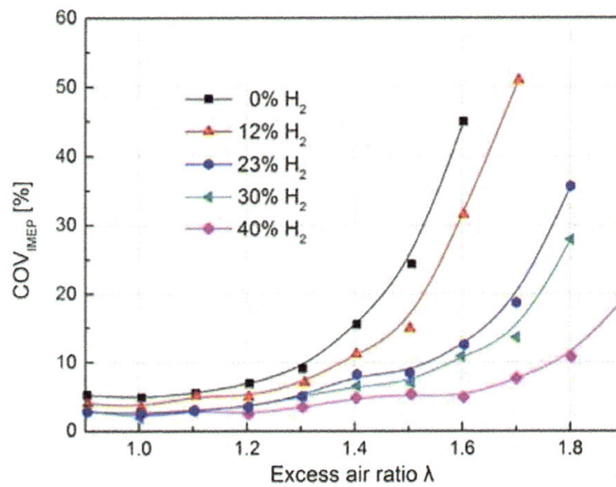

Fig. 9.14 Effect of λ and hydrogen enrichment of CNG on COV_{IMEP} (Wang et al. 2008)

Figure 9.14 shows the COV_{IMEP} versus λ for different H_2 fractions in HCNG mixtures. It can be observed from Fig. 9.14 that with an increase in λ, the COV_{IMEP} increased exponentially and it was maximum for 12HCNG at $\lambda = 1.7$. Wong and Karim (2000) analytically investigated the effect of hydrogen enrichment of CNG on the cyclic variations in a homogeneous charge compression ignition (HCCI) engine. They reported that increased H_2 fraction in HCNG was reduced the cyclic variations, which also extended the engine operating window. This was due to the fact that the addition of hydrogen to CNG increased the flame velocity of HCNG mixtures and reduced the CCVs.

9.3.2 HCNG Engine Emissions

Figure 9.15 shows the effect of excess air ratio on NO_x for different HCNG mixtures and compares it with baseline CNG. It was observed that NO_x emissions increased with increasing hydrogen fraction in the HCNG mixture at identical engine operating conditions. Mathai et al. (2012) reported 59.1% increase in average NO_x emissions for HCNG mixtures. Such increase in NO_x emission was primarily on account of elevated peak combustion temperature. It was observed that NO_x emission initially increased and attained maxima immediately after stoichiometry due to competing effects of temperature and oxygen availability and then decreased further with increasing λ.

Hoekstra et al. (1995) reported that NO_x emissions decreased with up to 30% hydrogen enrichment of CNG. Beyond this level of hydrogen fraction, no further reduction was seen. This was due to higher flame velocity and a consequently retarded spark advance for obtaining the maximum brake torque. However, increased combustion chamber temperature due to advance spark timing contributed to higher thermal NO_x generation.

Fig. 9.15 Variations of brake-specific NO_x emissions w.r.t. λ and varying H_2 fraction in HCNG mixtures (Ma et al. 2007)

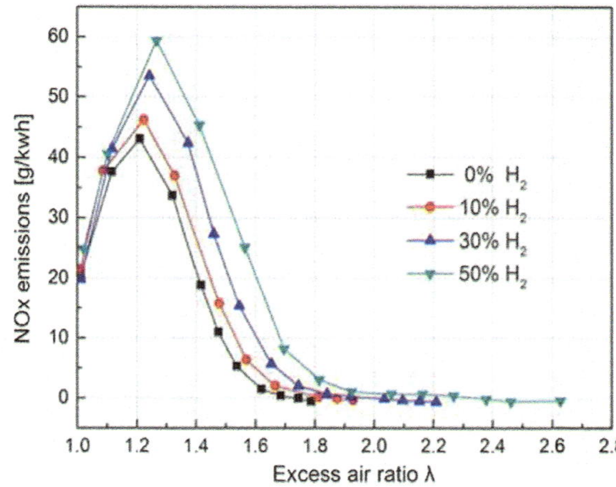

Fig. 9.16 Variations in brake-specific CO emission w.r.t. λ and varying hydrogen fraction in the HCNG mixture (Ma et al. 2007)

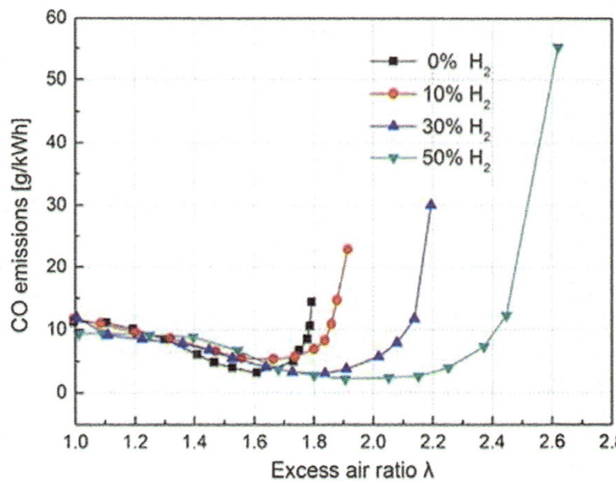

Figure 9.16 shows the variation of CO with λ for different HCNG mixtures. It was reported by several researchers that hydrogen enrichment of CNG reduced CO emission (Ceper et al. 2009; Deng et al. 2011; Park et al. 2011). De Simio et al. (2013) reported that CO emission reduced by $\sim$15.7% for 30HCNG and 16.2% for 40HCNG compared to baseline CNG. In another study, Mathai et al. (2012) performed an engine investigation for 60-h duration at 3000 rpm, 3.1 bar BMEP, and 24° CA bTDC spark timing and reported that CO emission reduced by 42.4% for 18HCNG compared to CNG.

Figure 9.17 shows the total particle number concentration variation with BMEP for various HCNG mixtures. Generally, the level of particulate number concentration is relatively lower for HCNG-fueled engines compared to conventional gasoline- or diesel-fueled engines. Limited research is available on the particulate emissions from HCNG engines in the open literature. Hora et al. (2016)

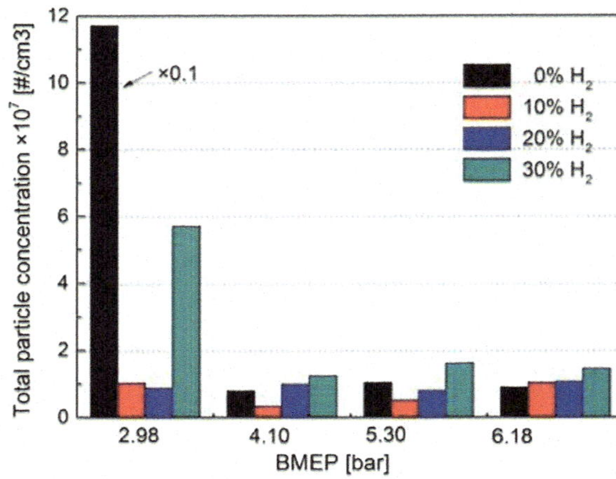

Fig. 9.17 Total particle number concentration w.r.t. BMEP for HCNG mixtures (Hora et al. 2016)

investigated the effect of hydrogen enrichment of CNG and found that it reduced the total particle numbers, especially at lower BMEPs. Particulate emissions from HCNG engines were mainly due to combustion of lubrication oil, which entered the combustion chamber through the passage in-between the cylinder liner and the piston rings (Hora et al. 2016; Singh et al. 2016). Figure 9.17 shows that at 2.98 bar BMEP, engine fueled with 10HCNG or 20HCNG mixtures emitted lower total particle number concentration compared to baseline CNG. At low BMEP, CNG-fueled engine emitted higher particle emissions because of incomplete combustion. Hydrogen enrichment results in higher flame velocity, advanced SoC, and shorter CD, thus leading to more complete combustion, which lowers the number of particulates emitted. In addition, the number of nano-sized particles emitted by the engine was observed to be much higher compared to that of particles in the accumulation mode (100–300 nm) for HCNG mixtures up to 30% hydrogen fraction. These results have critical consequences because the presence of nanoparticles in the atmosphere poses even higher risk to the human health than larger particles. Emission regulatory authorities worldwide are tirelessly working to bring legislation in all sectors of economy to limit both the mass and number of emitted particles from combustion sources, particularly from engines used for various applications, using variety of fuels. Therefore, time has come to pay greater attention on particle number emissions from HCNG-fueled engines as well.

9.4 Laser-Ignited Engine

9.4.1 Laser-Ignited Engine Combustion

Laser can be used to ignite an engine, as shown in Fig. 9.4. In Fig. 9.18, in-cylinder combustion parameters directly show the effect of test fuels on engine combustion.

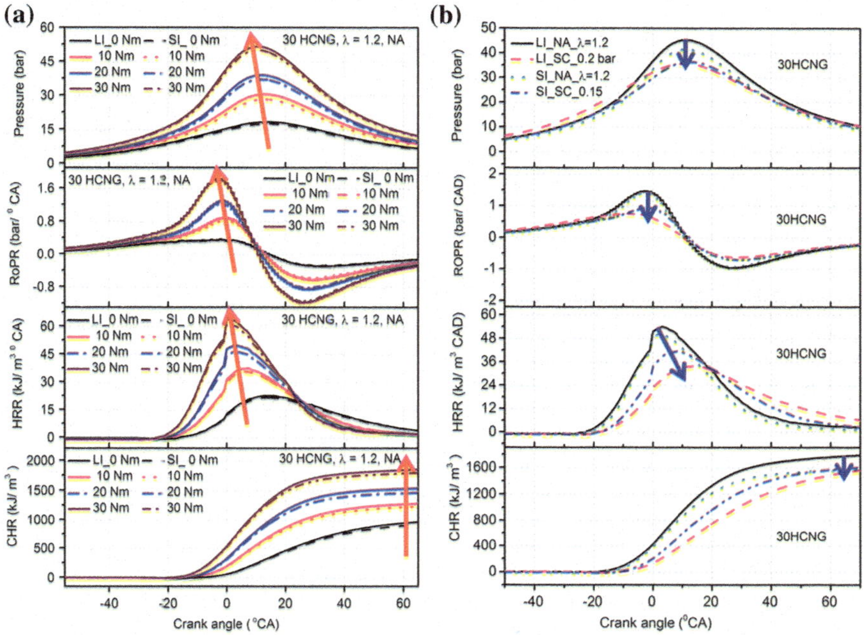

Fig. 9.18 Comparison of LI and SI modes of different HCNG–air mixtures for in-cylinder pressure, RoPR, HRR, and CHR at varying BMEP (Prasad 2018)

Combustion peak pressure (P_{max}) increased with increasing load and crank angle position of the peak pressure (ΘP_{max}) shifted toward the TDC (Fig. 9.18a). The shift in ΘP_{max} depends on chemical kinetics and flame velocity of the HCNG–air mixtures. P_{max} increased with increasing BMEP from no load to the full load because more fuel quantity was being injected at higher BMEP. Reaction kinetics accelerated for both higher BMEP and hydrogen fraction in the HCNG mixture.

Figure 9.18b shows the effect of boosting on the in-cylinder combustion parameters for 30HCNG. It can be observed that boosting reduces the peak in-cylinder pressure; however, pressure rise rate in LI mode was slightly higher than the SI mode. Increased hydrogen enrichment of CNG reduced the difference in peak in-cylinder pressure between NA and SC operating conditions. Engine torque output increased with increasing boost pressure initially, but it started reducing with further increasing boost. For 40 lpm fuel flow rate, 0.1 bar boost was found to be optimum boost pressure, at which the engine generated the maximum torque. HRR_{max} was 53.41 kJ/m^3.deg for CNG and 44.99 kJ/m^3.deg for 30HCNG. Different IMEP (4, 6, and 8 bar) under different engine speeds were also investigated. A comparison between conventional spark ignition and laser ignition with ignition energy of 30 and 60 mJ at turbulent as well as normal condition were compared.

Figure 9.19 depicts the combustion duration (in CAD) and ignition delay (in CAD) w.r.t. relative air–fuel ratio (λ) at 2000 rpm engine speed. There is

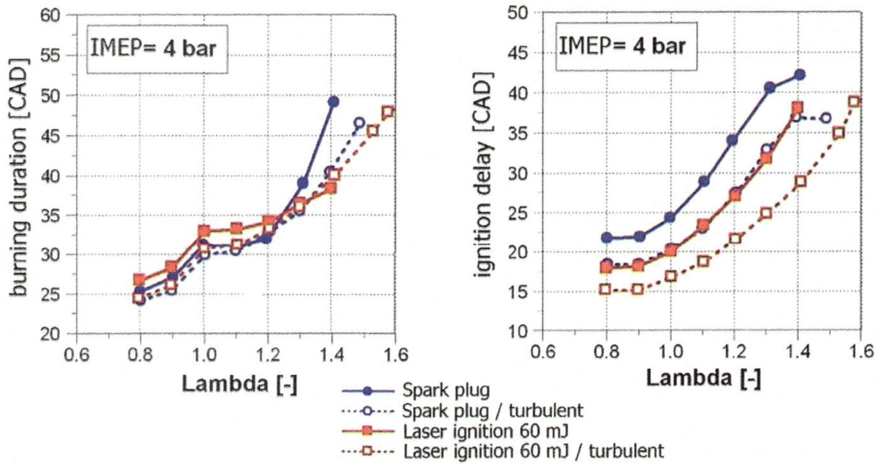

Fig. 9.19 Combustion duration and ignition delay w.r.t. λ for HCNG mixtures at 4 bar IMEP, 2000 rpm (Weinrotter 2006)

approximately identical combustion duration for λ up to 1.2, but further leaner mixtures showed shorter combustion duration. However, the ignition delay advanced by ~5 CAD in case of LI compared to SI.

9.4.2 *Laser-Ignited Engine Stability*

In Fig. 9.20, the COV_{IMEP} and ISFC are shown with varying λ. It was reported by the researchers that with LI, the λ window could be expanded far into the lean region. With turbulent combustion, it was possible to expand the λ window from 1.45 for SI to 1.65 for LI with the same COV_{IMEP}. The ISFC under LI was lower than the ISFC with SI, beyond $\lambda > 1.2$ for all other operating conditions.

9.4.3 *Laser-Ignited Engine Emissions*

Figure 9.21 depicts ISHC, $ISNO_x$ emissions, and the exhaust gas temperature (T_{exh}) w.r.t. λ. For leaner mixtures, T_{exh} reduced, and as expected, the thermal NO_x also reduced. If a maximum COV_{IMEP} was taken as 5% by definition, SI mode allowed a maximum λ of ~1.51; however, in LI mode, a lambda of ~1.6 was attained. With increased lean limit from 1.51 to 1.6, because of LI mode, NO_x emissions reduced by ~38%. Hence in comparison with reduced COV at the leanest mixtures, ISHC kept increasing with increasing λ because the combustion became more incomplete, and as a consequence, the unburned hydrocarbon emissions increased.

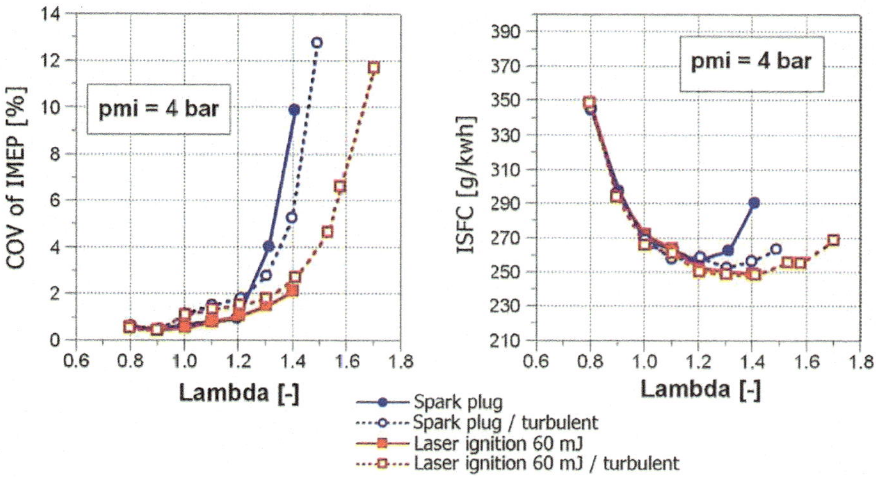

Fig. 9.20 COV_{IMEP} and ISFC w.r.t. λ for HCNG mixtures at 4 bar IMEP, 2000 rpm (Weinrotter 2006)

9.4.4 Effect of EGR on Laser-Ignited Engine

Besides the λ variations, EGR variation also affects the performance and combustion of a SI and LI engines. To understand the trend, variation in external un-cooled EGR was done to study its effect on combustion in SI and LI mode engine. Figure 9.22 illustrates the effect of EGR on combustion duration and ignition delay of SI and LI with 60 mJ pulse energy at 2000 rpm, at an IMEP of 4 bar under stoichiometric condition. For EGR > 10%, a slightly shorter combustion duration was observed in case of LI mode. The ignition delay was ~3–4 CAD shorter in case of LI over the entire range of EGR. Similar results were also observed for 1000 and 3000 rpm, and 2 and 6 bar IMEP.

9.4.5 Effect of Laser Ignition on PM Emissions

Singh et al. (2017) performed experiments and compared the particle number-size, surface area, and mass distributions using engine exhaust particle sizer (EEPS) of hydrogen-fueled engine in LI and SI modes at different BMEP. They reported that PM emissions increased with increasing engine load and LI mode exhibited relatively higher particle number concentration as well as particulate mass compared to SI mode; however, SI mode exhibited relatively larger count mean diameter (CMD). It was observed that fine particles with diameter less than 2.5 μm were more toxic compared to coarse particles (D_p < 10 μm) (Pope III et al. 1995). From

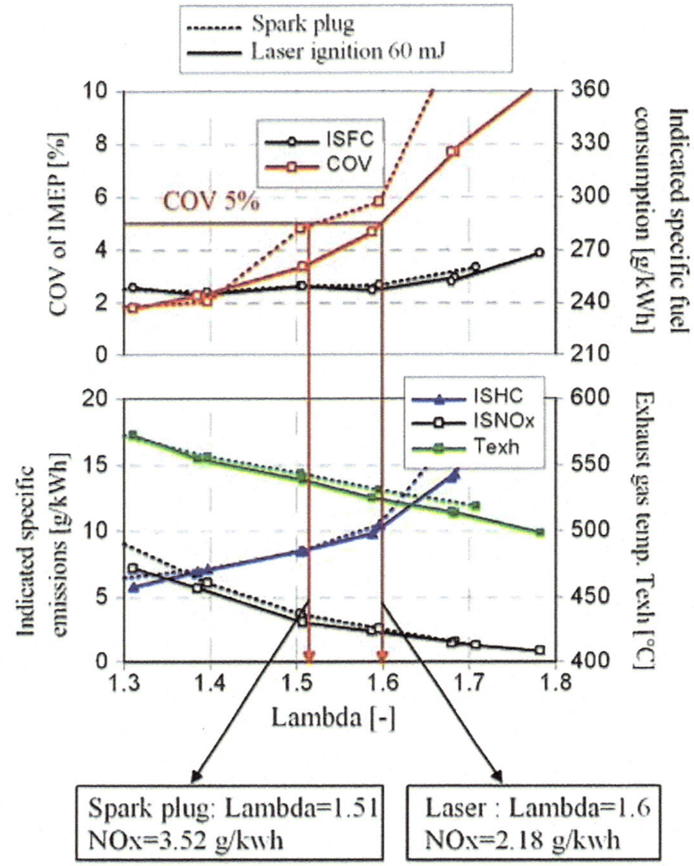

Fig. 9.21 COV_{IMEP}, ISFC, ISHC, $ISNO_x$ emissions, and T_{exh} w.r.t. λ for HCNG mixtures at 4 bar IMEP, 2000 rpm (Weinrotter 2006)

Fig. 9.23, it can be noted that for both LI and SI modes, nucleation mode particle (NMP) (i.e., particle diameter $D_p < 50$ nm) concentration was relatively higher compared to the accumulation mode particles (AMP) (i.e., particle diameter 50 nm $< D_p <$ 1000 nm) concentration.

LI system having intense plasma (∼2 order of magnitude higher temperature and ∼1 order of magnitude higher pressure) was compared with SI mode (Maly and Vogel 1979; Phuoc and White 1999). It was observed that peak AMP reduced with load in LI mode up to 15 nm size, while SI mode showed increased AMP up to 20 nm size. They also reported that particulate mass in engine exhaust increased in the same manner as the surface area distribution with increasing BMEP and was relatively higher for LI mode of combustion. Particulate mass was calculated from the particulate number-size distribution, considering particles to be spherical with

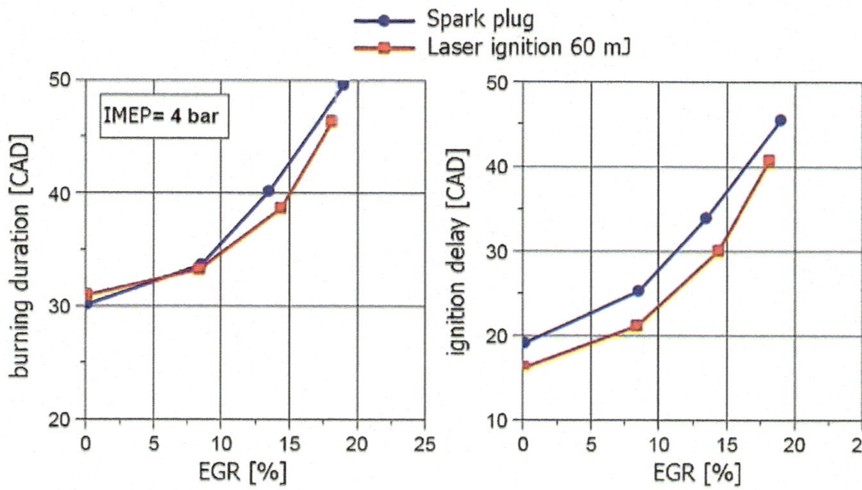

Fig. 9.22 Effect of EGR on burning duration and ignition delay at λ = 1.0 for HCNG mixtures at 4 bar IMEP, 2000 rpm (Weinrotter 2006)

identical density (1 g/cm^3) (Engine Exhaust Particle Sizer™ 2009; Gupta et al. 2010). It was observed from Fig. 9.23 that nuclear mode particles were ~70% of total but their contribution was significantly lower in the mass%. Condensation of organic species on primary particles resulted in the formation of accumulation mode particles.

9.4.6 *Effect of Plasma Location on Laser-Ignited Engine*

Weinrotter et al. (2004) investigated effect of plasma position in the combustion chamber on the engine performance and emissions (Fig. 9.24). It was observed that these parameters were optimum at a typical plasma position (12 mm from the cylinder head) without EGR at 2000 rpm engine speed, IMEP of 8 bar, and λ = 1.0.

9.5 Conclusions

The engine combustion, performance, and emissions were characterized for different relative air–fuel ratio (λ) for different HCNG mixtures in LI and SI modes. From the naturally aspirated (NA) and supercharged engine experiments, the following conclusions were drawn:

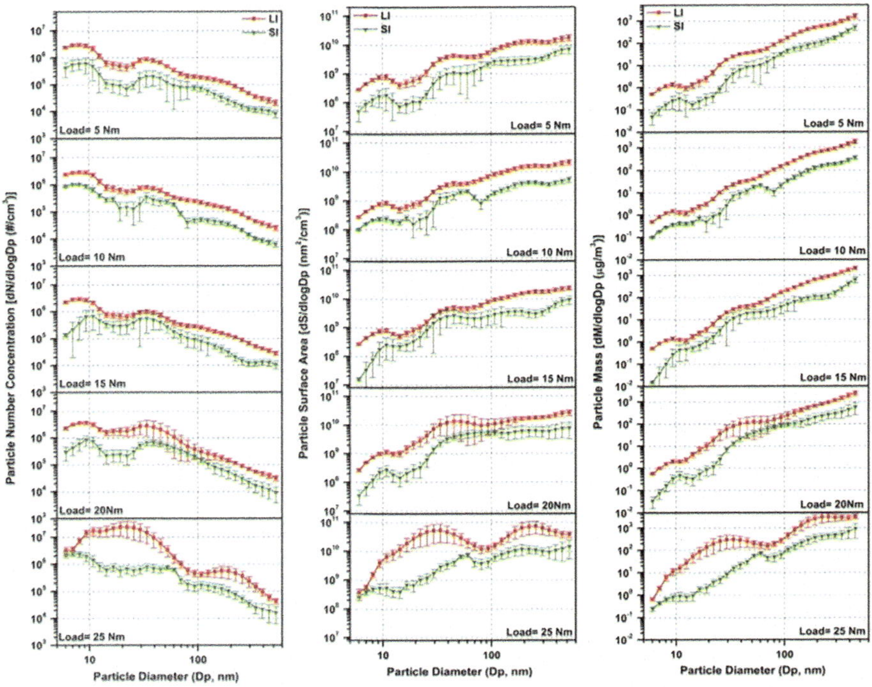

Fig. 9.23 Comparison of particle number-size, surface area, and mass distribution from H_2 under LI and SI systems at different BMEP (Singh et al. 2017)

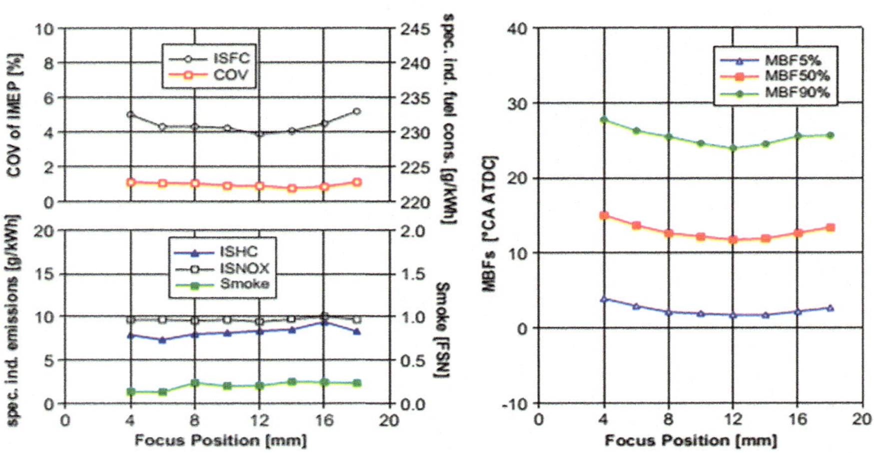

Fig. 9.24 Variation of the focus position at $n = 2000$ min − 1, IMEP = 8 bar, homogeneous mode, $\lambda = 1.0$, w/o EGR (Weinrotter et al. 2004)

- SoC, combustion phasing, and EoC advanced with increasing BMEP in SI mode for all HCNG mixtures.
- COV_{IMEP} reduced with increasing hydrogen enrichment of CNG.
- $BSNO_x$ emissions increased with increasing BMEP but reduced with increasing λ.
- LI increased the lean-burn limit for all test fuels compared to conventional SI mode.
- The engine torque increased with increasing boost, and maximum torque was observed at optimum boost of 0.1 bar.
- Boost reduced P_{max}, RoPR, HRR, and BTE; however, BSFC and BSEC improved for both LI and SI modes for all HCNG mixtures. These parameters were relatively even higher at maximum possible boost for LI.

In summary, laser-ignited HCNG-fueled engine under lean condition leads to higher engine power output, higher BTE, reduced BSFC, and lower $BSNO_x$ emissions. Laser ignition of HCNG mixture reduced GHG emissions, and NO_x significantly, extended the lean-burn limit, improved the BTE, and produced higher specific power output. Such an engine can be potentially used in important applications such as automotive, military, and stationary power plants.

References

Bauer H (1998) Ottomotor-management. Robert Bosch, GmbH

Bysveen M (2007) Engine characteristics of emissions and performance using mixtures of natural gas and hydrogen. Energy 32:482–489

Ceper BA, Akansu SO, Kahraman N (2009) Investigation of cylinder pressure for H_2/CH_4 mixtures at different loads. Int J Hydrogen Energy 34:4855–4861

Dale JD, Smy PR, Clements RM (1978) Laser ignited internal combustion engine—an experimental study. SAE technical paper, p 780329

De Simio L, Gambino M, Iannaccone S (2013) Experimental and numerical study of hydrogen addition in a natural gas heavy duty engine for a bus vehicle. Int J Hydrogen Energy 38:6865–6873

Deng J, Ma F, Li S, He Y, Wang M, Jiang L, Zhao S (2011) Experimental study on combustion and emission characteristics of a hydrogen-enriched compressed natural gas engine under idling condition. Int J Hydrogen Energy 36:13150–13157

EIA U (2006) In: International energy annual 2006. Washington. http://www.eia.doe.gov.iea

Engine Exhaust Particle Sizer™ (2009) Spectrometer model 3090. Operation and Service Manual, TSI, USA

Fayaz H, Saidur R, Razali N, Anuar FS, Saleman AR, Islam MR (2012) An overview of hydrogen as a vehicle fuel. Renew Sustain Energy Rev 16(8):5511–5528

Gupta T, Kothari A, Srivastava DK, Agarwal AK (2010) Measurement of number and size distribution of particles emitted from a mid-sized transportation multipoint port fuel injection gasoline engine. Fuel 89(9):2230–2233

Helmreich HW, Lochner S. Natural gas in road transportation—a low emission bridging technology? EWI working paper 2011. (http://www.ewi.unikoeln.de/fileadmin/user_upload/Publikationen/Working_Paper/EWI_WP_11-14_Natural_gas_road_transportation.pdf)

Hoekstra RL, Collier K, Mulligan N, Chew L (1995) Experimental study of a clean burning vehicle fuel. Int J Hydrogen Energy 20:737–745
Hora TS, Shukla PC, Agarwal AK (2016) Particulate emissions from hydrogen enriched compressed natural gas engine. Fuel 166:574–580
Jones FL (1963) The mechanism of electrode erosion in electrical discharge. Platin Met Rev 7 (2):58–65
Karim GA (2003) Hydrogen as a spark ignition engine fuel. Int J Hydrogen Energy 28:569–577
Karim GA, Wierzba I, Al-Alousi Y (1996) Methane–hydrogen mixtures as fuels. Int J Hydrogen Energy 21(7):625–631
Karner D, Brayer R, Peterson D, Kirkpatrick M, Francfort J, Kirpatrick M (2003) Freedom car and vehicle technologies program–advanced vehicle testing activity–Arizona public service-Alternative fuel (Hydrogen) pilot plant
Kofler H (2007) Corona and laser ignition in internal combustion engines—a comparison to conventional spark plug ignition. Master dissertation, Vienna University of Technology
Lim G, Lee S, Park C, Choi Y, Kim C (2013) Effects of compression ratio on performance and emission characteristics of heavy-duty SI engine fuelled with HCNG. Int J Hydrogen Energy 38:4831–4838
Ma F, Wang Y, Liu H, Li Y, Wang J, Zhao S (2007) Experimental study on thermal efficiency and emission characteristics of a lean burn hydrogen enriched natural gas engine. Int J Hydrogen Energy 32:5067–5075
Ma F, Liu H, Wang Y, Li Y, Wang J, Zhao S (2008) Combustion and emission characteristic of a port injection HCNG engine under various ignition timings. Int J Hydrogen Energy 33:816–822
Ma F, Wang M, Jiang L, Chen R, Deng J, Naeve N, Zhao S (2010) Performance and emission characteristics of a turbocharged CNG engine fueled by hydrogen-enriched compressed natural gas with high hydrogen ratio. Int J Hydrogen Energy 35(12):6438–6447
Ma F, Li S, Zhao J, Qi Z, Deng J, Naeve N, Zhoa S (2012) Effect of compression ratio and spark timing on the power performance and combustion characteristics of an HCNG engine. Int J Hydrogen Energy 37:18486–18491
Maly R, Vogel M (1979) Initiation and propagation of flame fronts in lean CH4-air mixtures by the three modes of the ignition spark. In: Symposium (international) on combustion, pp 821–831
Mathai R, Malhotra RK, Subramanian KA, Das LM (2012) Comparative evaluation of performance, emission, lubricant and deposit characteristics of spark ignition engine fueled with CNG and 18% hydrogen-CNG. Int J Hydrogen Energy 37:6893–6900
McNeil DH (2005) Minimum ignition energy for laser spark ignition. Proc Combust Inst 30 (2):2913–2920
Park C, Kim C, Choi Y, Won S, Moriyoshi Y (2011) The influences of hydrogen on the performance and emission characteristics of a heavy duty natural gas engine. Int J Hydrogen Energy 36:3739–3745
Park C, Kim C, Choi Y (2012) Power output characteristics of hydrogen-natural gas blend fuel engine at different compression ratios. Int J Hydrogen Energy 37:8681–8687
Phuoc TX (2000) Laser spark ignition: experimental determination of laser induced breakdown thresholds of combustion gases. Opt Commun 175:419–423
Phuoc TX, White FP (1999) Laser-induced spark ignition of CH4/air mixtures. Combust Flame 119(3):203–216
Pope III CA, Thun MJ, Namboodiri MM, Dockery DW, Evans JS, Speizer FE et al (1995) Particulate air pollution as a predictor of mortality in a prospective study of U.S. adults. Am J Respir Crit Care Med 151:669–674
Prasad RK (2018) Laser plasma ignited hydrogen enriched compressed natural gas engine development and experimental evaluation. Doctoral dissertation, Indian Institute of Technology Kanpur
Raizer YP (1997) Gas discharge physics. Springer, Berlin
Raman V, Hansel J, Fulton J, Lynch F (1994) Hythane—an ultraclean transportation fuel. Hydrogen Energy Prog 3:1797

Shudo T, Shimamura K, Nakajima Y (2000) Combustion and emissions in a methane DI stratified charge engine with hydrogen pre-mixing. JSAE Rev 21:3–7

Singh AP, Pal A, Agarwal AK (2016) Comparative particulate characteristics of hydrogen, CNG, HCNG, gasoline and diesel fueled engines. Fuel 185:491–499

Singh AP, Pal A, Gupta NK, Agarwal AK (2017) Particulate emissions from laser ignited and spark ignited hydrogen fueled engines. Int J Hydrogen Energy 42(24):15956–15965

Sorensen B (2011) Hydrogen and fuel cells: emerging technologies and applications. Academic Press, Oxford

Taira T (2015) Overview of optical/laser technological advances leading to practical laser ignition systems. In: Laser ignition conferenc, T3A.2

Tangöz S, Akansu SO, Kahraman N, Malkoç Y (2015) Effects of compression ratio on performance and emissions of a modified diesel engine fueled by HCNG. Int J Hydrogen Energy 40:15374–15380

The outlook for energy: a view to 2040. http://cdn.exxonmobil.com/ ~ /media/global/files/outlook-for-energy/2016/2016outlook-for-energy.pdf

Vander Hoeven M (2015) Energy and climate change—world energy outlook special report. International Energy Agency. https://www.iea.org/publications/freepublications/publication/WEO2015SpecialReportonEnergyandClimateChange.pdf

Verma G, Prasad RK, Agarwal R, Jain S, Agarwal AK (2017) Experimental investigations of combustion, performance and emission characteristics of a hydrogen enriched natural gas fuelled prototype spark ignition engine. Fuel 190:318–327

Wang J, Chen H, Liu B, Huang Z (2008) Study of cycle-by-cycle variations of a spark ignition engine fueled with natural gas–hydrogen blends. Int J Hydrogen Energy 33:4876–4883

Weinrotter M (2005) Laser ignition of internal combustion engines—basic laser and ignition optics developments, engine application and optical diagnostics. Doctoral dissertation, Vienna University of Technology

Weinrotter M, Kopecek H, Wintner E (2004) Laser ignition, optics and contamination of optics in an I.C. engine. In: ASME ICED fall technical conference proceeding, Paper No. ICEF 2004-833, pp 11–17

Wintner E, Kofler H (2015) Laser ignition for more than four decades: solutions and current problems. Laser ignition conference, T5A.1

Wong YK, Karim GA (2000) An analytical examination of the effects of hydrogen addition on cyclic variations in homogeneously charged compression-ignition engines. Int J Hydrogen Energy 25(12):1217–1224

Yalin AP, Reynolds AR, Joshi S, Defoort MW, Willson B, Matsuura Y, Miyagi M (2006) Development of a fiber delivered laser ignition system for natural gas engines. In: ASME internal combustion engine division spring technical conference proceeding, pp 117–122

Chapter 10
Particulate Matter and Its Impact on Human Health in Urban Settings

Dev Prakash Satsangi and Avinash Kumar Agarwal

Abstract In the current era of development, all living beings on earth are confronting a significant increase in diseases related to airways and lungs. Humans are introduced to new diseases such as asthma, chronic obstructive pulmonary disease (COPD), lung cancer. These diseases are essentially caused by the inhalation of particulate matter from various combustion sources, including internal combustion engines. The situation is getting worse, and it is affecting all living being in spite of their financial status since everyone needs to breathe for survival. The severity of these diseases depends upon the dose of particulates inhaled by humans. Particulate matter concentration in the environment is accelerating due to the increasing number of different kind of vehicles worldwide using petroleum-based hydrocarbon fuels, which is increasing due to increasing global prosperity. Therefore, an urgent switch from the commercially available energy sources to carbon-free fuel economy is necessary for a clean, healthy, and livable environment.

Keywords Diesel engines · Particulate matter · Respiratory diseases

Abbreviations

COPD	Chronic obstructive pulmonary disease
CRDI	Common rail direct injection
DNA	Deoxyribonucleic acid
DOC	Diesel oxidation catalysts
DPF	Diesel particulate filter
IC Engines	Internal combustion engines
IT	Interim target
PAH	Polycyclic aromatic hydrocarbons
PM	Particulate matter
PME	Palm methyl ester

D. P. Satsangi · A. K. Agarwal (✉)
Indian Institute of Technology Kanpur, Kanpur 208016, India
e-mail: akag@iik.ac.in

A. K. Agarwal et al. (eds.), *Methanol and the Alternate Fuel Economy*,
Energy, Environment, and Sustainability,
https://doi.org/10.1007/978-981-13-3287-6_10

ROS	Reactive oxygen species
SEM	Scanning electron microscope
WHO	World Health Organization

10.1 Introduction

Worldwide, many studies have indicated that the air pollution affects both humans and the environment significantly. Air pollution is a complex mixture of fine particles and gases, and the concentration of these contents varies in space and time both. Average human health is progressively worsening, and the problems range from difficulty in breathing to the diagnosis of fatal heart and lung diseases, all attributed to these air pollutants. Particulates, which exist in the environment, are the deadliest part of air pollution because they can penetrate deeply into the lungs and the bloodstream. The atmospheric particulate matter is sometimes also referred to as particulate matter (PM) or particulates. Particulate matter is a mixture of solid particles and liquid droplets, which are very small in size and remain suspended in the atmospheric air. Typically, PM is classified by their size, composition, and origin (https://www.conserve-energy-future.com/causes-and-effects-of-particulate-matter.php). In many countries, especially the developing ones, the level of $PM_{2.5}$ and PM_{10} exceed in comparison with guidelines or standards specified by the world health organization (WHO). PM is a critical component of air pollution, which has significant adverse effects on the human health and the environment. In the last 2–3 decades, the open literature on the atmospheric PM or aerosols has increased exponentially.

Different developing countries in the world are leading to excessive use of energy and quantum jump in construction activities. These economic activities are very important sources of environmental particulates such as road dust, industries, fossil fuel combustion, and electricity generation using coal. In the context of engines, PM emission is the most negative aspect of the internal combustion (IC) engines used for a large variety of applications in everyday life, including transport and stationary power generation.

State of global air report 2018 indicated that worldwide exposure to air pollution and PM had increased enormously over the years. Hence, the diseases due to these air pollutants have also grown exponentially worldwide. PM, especially $PM_{2.5}$, is reported to be the sixth highest risk factor responsible for the early deaths of humans (Fig. 10.1). In the year 2016, nearly 4.1 million people died due to lung cancer, respiratory infections, chronic lung diseases, heart diseases, and stroke (Health Effects Institute 2018). Deaths of these people were correlated more to relatively higher exposure to $PM_{2.5}$ compared to other factors such as the use of alcohol, low physical activity, or high sodium intake. Another essential component of ambient pollution is ground-level ozone, which is an outdoor air pollution

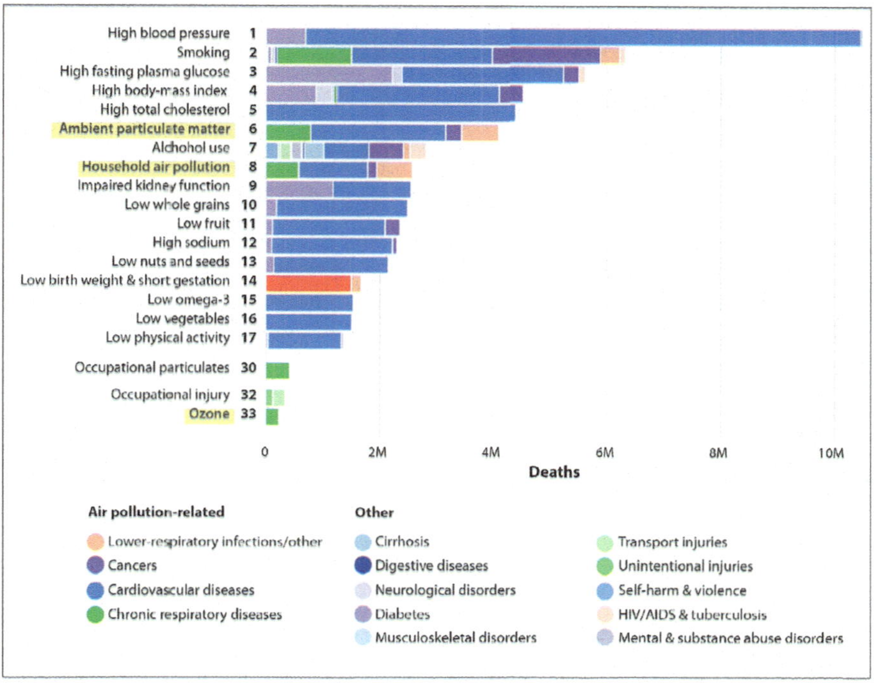

Fig. 10.1 Global health risk factors and their respective rankings, responsible for the early deaths of both males/females in the year 2016 (Health Effects Institute 2018)

component. Ozone was responsible for nearly 234,000 deaths due to chronic lung diseases in the year 2016. Also, there are household air pollution factors, such as the use of solid fuels (coal, wood, dung, etc.), which are responsible for substantial issues in people's health (Health Effects Institute 2018).

The emission guidelines established by WHO has been tightened due to a rapid increase in $PM_{2.5}$ levels in the ambient air. The air quality guidelines by WHO on the annual average basis for the $PM_{2.5}$ are set to be 10 $\mu g/m^3$. Also, for many countries, which have a high level of air pollution, WHO has set three interim target limits (35, 25 and 15 $\mu g/m^3$) for $PM_{2.5}$. It has been reported by WHO that the 95% of the global population is living in the regions, which violate $PM_{2.5}$ guidelines (Health Effects Institute 2018). WHO has divided the world into different interim target (IT) areas. The report concluded that 58% of the global population is living in the region, which exceeded IT-1, i.e., 35 $\mu g/m^3$. Similarly, 69 and 85% global population is living in the areas, which exceeded IT-2, i.e., 25 $\mu g/m^3$ and IT-3, i.e., 15 $\mu g/m^3$, respectively (Health Effects Institute 2018). Figure 10.2 shows the trends of a population-weighted annual average of $PM_{2.5}$ concentration of different countries between 2010 and 2016. Also, the global weighted average of $PM_{2.5}$ was 43.2 $\mu g/m^3$ for 2010 compared to 51.1 $\mu g/m^3$ for 2016, which shows that the PM emission level increased by $\sim$18%.

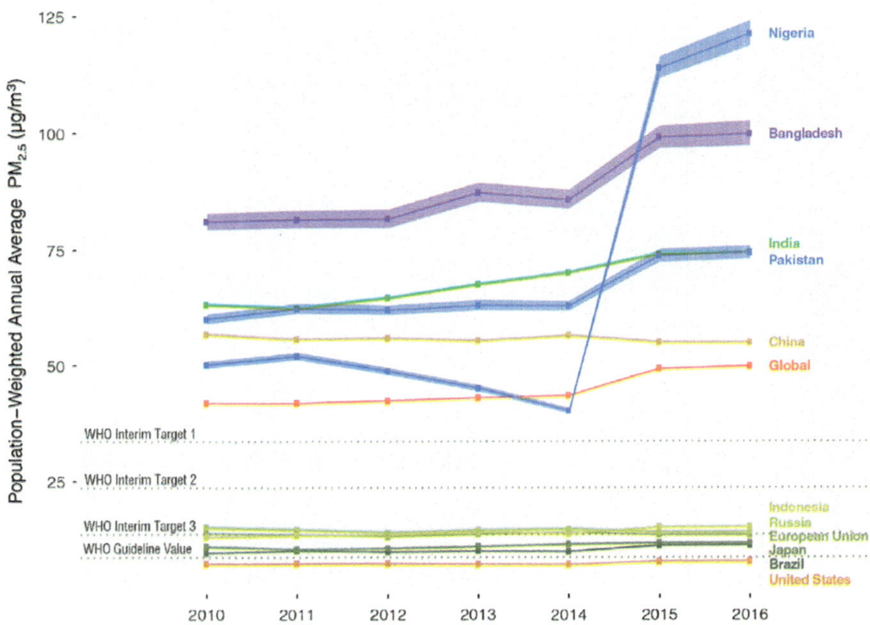

Fig. 10.2 Different countries and their population-weighted annual average of $PM_{2.5}$ for 2010–2016 (Health Effects Institute 2018)

Research findings are expected to be useful for advanced research on PM. These research findings on the PM play an essential role in characterizing the PM, the health effects of PM exposure and the underlying physics of these effects. Therefore, further studies can be instrumental in developing technology and methods for reducing the PM emissions from various sources, including IC engines. Establishing a link between the sources of PM, and the human exposure, and finally with various respiratory diseases can be very handy. Research on particulates requires the involvement of multiple domain experts such as engineers, scientists, exposure assessors, toxicologists, epidemiologists, and clinicians (National Research Council 2004).

10.2 Particular Matter and Formation Process

Particulate matter (PM) is made up of soot and other materials deposited on the surface of the soot particles. Soot particles are mostly formed during combustion of fuel-rich zones in the combustion chamber, which have insufficient oxygen in the presence of high temperature. Afterward, hydrocarbons condense and deposit on the surface of the carbonaceous core to form the particulates. Figure 10.3 shows a schematic representation of the PM emitted from IC engines.

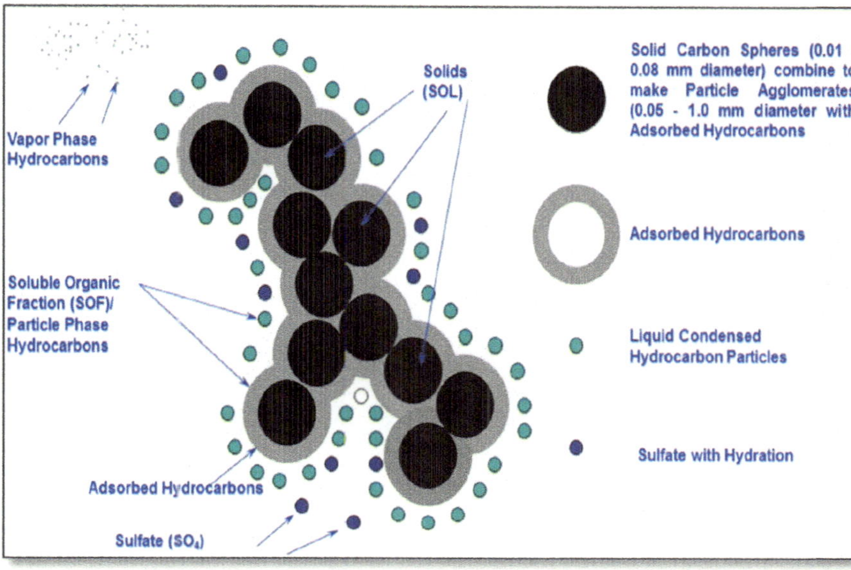

Fig. 10.3 Schematic representation of PM emitted by IC engines (Mohankumar and Senthilkumar 2017)

In the ambient air, particles like dust and smoke are somewhat bigger and dark enough to be visible. However, for viewing the small-sized particles, scanning electron microscope (SEM) is used. PM can be classified into PM_{10} and $PM_{2.5}$ in the context of size. PM_{10} is the group of tiny particles, which can be inhalable and have their diameter less than 10 μm.

Similarly, $PM_{2.5}$ represents the group of very fine particles, which can be inhaled. The diameter of the $PM_{2.5}$ particles is less than 2.5 μm (https://www.epa.gov/pm-pollution/particulate-matter-pm-basics). Figure 10.4 shows the size comparison of PM_{10} and $PM_{2.5}$. One can note that the size of PM is at least five times smaller than the width of a human hair. PM with different sizes behave differently in the atmosphere. Therefore, PM is classified based on their aerodynamic properties. The reasons behind these classifications are (http://www.thecmmgroup.com/know-thy-enemy-particulate-matter/):

- Aerodynamic properties of PM help in their transit and elimination from the atmospheric air.
- Aerodynamic properties of PM govern their deposition behavior within the respiratory system.
- Aerodynamic properties of PM are affected by their chemical composition and their source of origin.

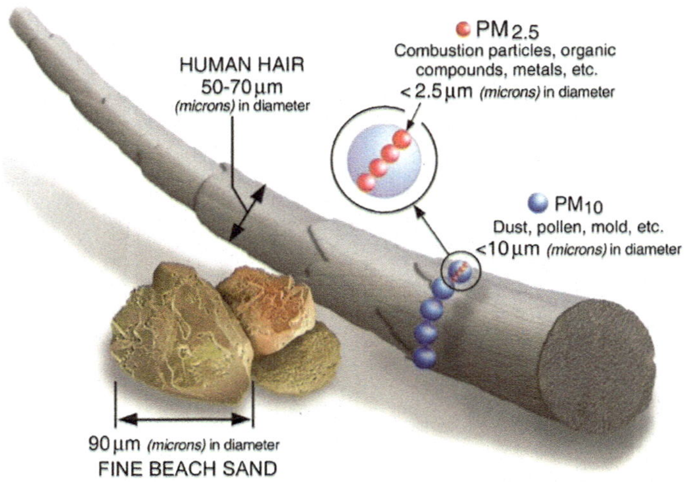

Fig. 10.4 Size comparison of particulates (Accessed on 09 July 2018)

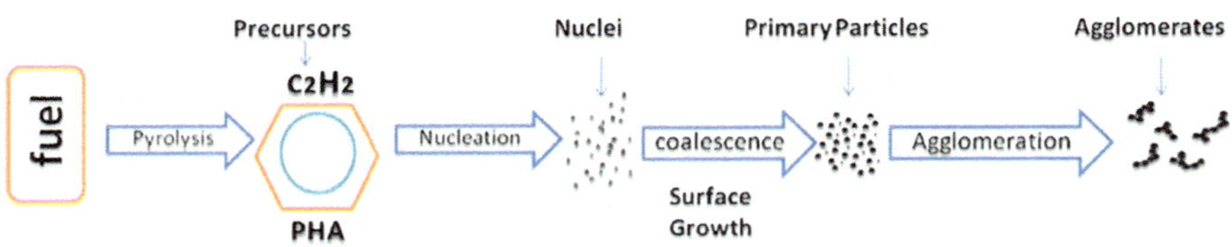

Fig. 10.5 Steps followed in the evolution of PM (Mohankumar and Senthilkumar 2017)

The constituents of PM can be soluble or insoluble. The soluble parts of PM include lubricating oil, partially oxidized fuel, oil, alkanes, alkenes, aldehydes, aliphatic hydrocarbons, polycyclic aromatic hydrocarbons and its derivatives (Williams et al. 1987). The conversion of liquid fuel to gaseous phase and solid soot particles and then to mature particulate formation involves a process of different intermediate stages namely (i) pyrolysis, (ii) nucleation, (iii) surface growth, (iv) coalescence, (v) agglomeration, and (vi) oxidation (Mohankumar and Senthilkumar 2017). Figure 10.5 shows the size comparison of particulates with other substances.

The physical process of soot formation is controlled by various vital factors namely the chemical composition of liquid fuel, in-cylinder temperature, pressure, and fuel injection strategies during the combustion process. Also, the formation of soot and oxidation process rate is mainly affected by the in-cylinder pressure and temperature (Hauser et al. 2001). A detailed explanation of these processes is given in the following sections.

10.2.1 Pyrolysis

In the pyrolysis process, the molecular structure of fuel, which is an organic compound, undergoes various chemical changes in the presence of high temperature and deficient oxygen (Smith 1981). The pyrolysis process generated soot precursors. The rate of pyrolysis and oxidation process is influenced by the type of flame produced during combustion. Formation of soot was observed to be higher in premixed flames and relatively lower in the diffusion flames. It is because the concentration of oxygen is relatively higher for premixed flames compared to the diffusion flames. Also, the oxidation rate increases with increasing temperature. Hence, the formation of soot is mainly influenced by the temperature and local oxygen content. Also, the report suggested that the existence of radicals like O, O_2, and OH triggers pyrolysis. The products of pyrolysis process are polycyclic aromatic hydrocarbons (PAH's), polyacetylenes, and unsaturated hydrocarbons (Mohankumar and Senthilkumar 2017). Among these products of pyrolysis, acetylene plays an important role. In the fuel, if sufficient amount of OH and O radicals exist then during the oxidation process, inert products are formed from acetylene (Smith 1981).

10.2.2 Nucleation

In the process of nucleation, gas phase reactants get converted into tiny particles. The diameter of nuclei formed in the luminous flames varies from 1.5 to 2 nm (Haynes et al. 1991). Nucleation process takes place when the temperature of combustion reaches in the range of 1300–1600 K. Formation of large-sized aromatic molecules occurs by radical addition of small size hydrocarbons during the particle inception process. The nuclei of the particles formed during the process do not contribute to the total mass of the soot. In fact, these nuclei act as a site for surface growth. In regions of high temperature and high ion concentration, a reduction in the premixed and diffusion flame nucleation has been observed (Mohankumar and Senthilkumar 2017). Bruce et al. (1991) stated that the process of nucleation has the following crucial steps.

- In the first step, chain molecules of acetylene get converted into a ring-type structure (benzene) by cyclization process.
- In the next step, these aromatic rings in the presence of lower temperature are dehydrogenated and turned into polycyclic compounds.
- In the last step, separation and re-cyclization of aromatic rings occur at high temperature.

10.2.3 Surface Growth

A large number of small particles are produced after completion of nucleation. The size of PM increases during the process of surface growth and therefore is responsible for an increase in the soot mass. After that, gas phase hydrocarbons deposit on the surface of these spherules to form the nuclei. It was observed that acetylene increases the soot mass, without changing the number of particles. This process continues although the particles shift from the highly reactive regions to less reactive regions. These regions possess a lower temperature and a relatively lower number of hydrocarbons for the soot initiation (Haynes and Wagner 1981). The rate of soot formation during the surface growth process primarily depends upon the numbers of nuclei (Smith 1981). After the formation of nuclei, surface growth process may last for a maximum of 0.05 ms. Therefore, the mass and volume fraction of soot formation is mainly affected by the residence time of the surface growth process. The rate of surface growth is lower for larger particles, compared to the smaller particles (Haynes and Wagner 1981).

10.2.4 Coalescence and Agglomeration

Coalescence and agglomeration is the process of particle combination during combustion. Particle collision occurring during the formation stage is termed as 'agglomeration'. Agglomeration leads to the decay in the number of particles, however, their size increases. In both the processes (coalescence and agglomeration), the spherical particles collide to form a single spheroid. These processes are affected by engine operating conditions, which finally decide the size of the primary spherical particles. Typically, spherical particles may vary in the range of 20–70 nm (Mohankumar and Senthilkumar 2017). Bruce et al. (1991) and Lee et al. (2001) studied the size of the primary particles using sampling probe and optical scattering technique and reported the size of these particles in the order of 30–70 nm. In other studies, by Tree and Foster (1994), and Pinson et al. (1994), the size of primary particles size was reported to be 30–50 and 40–65 nm, respectively. In the next stage, the agglomerated process occurs to form a chain-like structure, whose size is in the range of 100 nm to 2 μm (Mohankumar and Senthilkumar 2017).

10.2.5 Oxidation

Next, in the oxidation process, molecules such as carbon and hydrocarbon oxidize. The carbon molecules do not play a role in the formation of soot due to its partial oxidation into CO. Also, the oxidation process does not influence the process of surface growth and coagulation (Mohankumar and Senthilkumar 2017).

10.3 Particular Matter and the Human Respiratory System

Many epidemiological studies in the world have reported that the human exposure to the particulates is responsible for the increase in respiratory symptoms and a reduction in pulmonary functions. Hence, an increase in the rate of hospitalization, as well as in mortality rates, has been observed due to pulmonary infections and respiratory diseases both (Jang 2012; WHO 2003). Therefore, it is necessary to reduce particle concentration in the ambient air. Figure 10.6 shows a pyramid representation, which indicates the severity of human health when exposed to air pollutants. It indicates that if a person is exposed to air pollution for a long duration, then it may lead to causality. It has been reported by WHO that globally, ~24% of respiratory diseases and ~23% of deaths due to these respiratory diseases are due to environmental factors (Pruss-Ustun and Corvalán 2006). Also, the PM emission emanating from various combustion sources has increased the possibility of allergic diseases, especially in the urban areas.

Figure 10.7 shows particulate penetration intensity in the respiratory tract. This figure reveals that how far these particles of different sizes can reach into the human body, lungs, and bloodstream (http://www.thecmmgroup.com/know-thy-enemy-particulate-matter/).

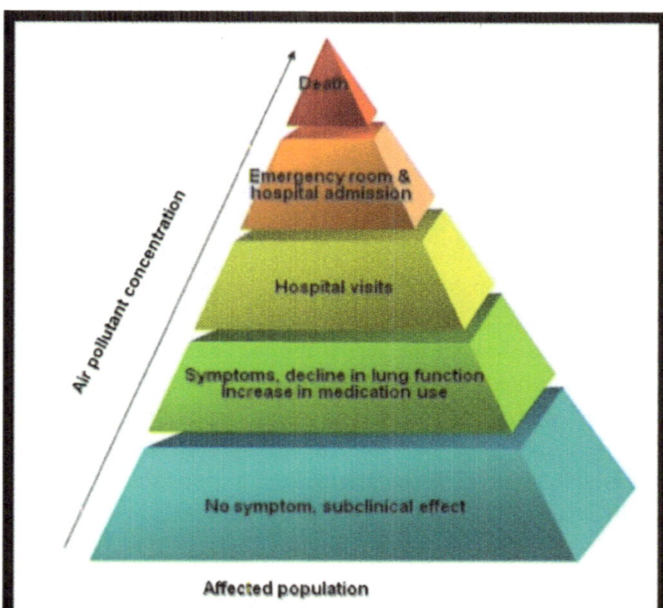

Fig. 10.6 Severity pyramid of human health due to exposure to air pollutants (Jang 2012)

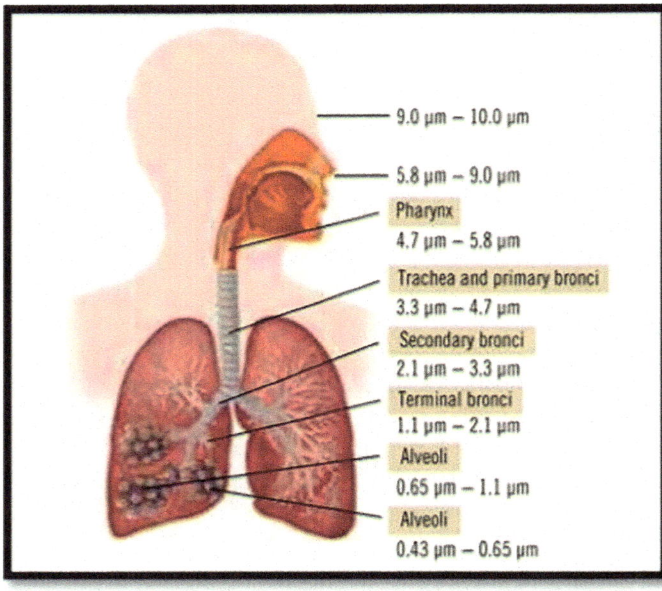

Fig. 10.7 Particle penetration in the human respiratory system, depending on size (http://www.thecmmgroup.com/know-thy-enemy-particulate-matter/ 2018)

The coarse particles (PM_{10}) have a more localized effect due to their size; hence, they are unable to travel far and wide into the respiratory system. Inhaling high concentration of PM_{10} can lead to reactions such as coughing and irritation. The size of PM_{10} makes them not to reach gas-exchange zone deep into the human lungs. On the other hand, fine particles ($PM_{2.5}$) are smaller in size and can travel into the respiratory system deeper. Hence, particulates, especially $PM_{2.5}$, are considered to be much more dangerous compared to PM_{10}. Therefore, $PM_{2.5}$ has a greater tendency to damage the respiratory system. The size of the PM and the efficiency of airway defense mechanism drives how deep these particles can penetrate in the trachea-bronchial tract (D'Amato et al. 2010). Particles with size less than 10 µm are contained/arrested in the large upper branches, just below the throat. Particles with a size less than 5 µm pass through the top of the lungs and move into the bronchial tubes. Particles smaller than 2.5 µm are the most dangerous because they can reach the lungs (i.e., alveolar region). Fine particles may be soluble or insoluble in water. The soluble particles pass, therefore, directly into the blood (alveolar capillaries); however, the insoluble particles remain trapped deep within the lungs for a long time (Jang 2012). Inhaling PM in high amounts, therefore, is responsible for dysfunctional lungs and heart.

10.4 Toxic Effects of Particulates on Human Health

Particulates have become a health burden due to its harmful effects on the human health. In the year 2015, deaths reported due to the toxic nature of PM increased by 7.8%, compared to 2005. PM acquired sixth position in a list of 10 most dangerous factors responsible for global disability (Forouzanfar et al. 2016). Since PM can directly interact with the human body, hence it can be detected in various human organs such as liver, lungs. But, the deposition of PM is primarily detected in the lungs (Aalapati et al. 2014). No standard dose has been defined so far for PM exposure due to complex nature of particulates and their inconsistent chemical and physical properties.

Once the chemical action of PM damages the lungs, then the recovery of damaged lungs to their original stage is impossible (Kim et al. 2017). Kim et al. (2017) reported that the exposure to PM could lead to oxidative stress and inflammation of breathing organs. Therefore, early diagnosis of lung diseases is essential for obtaining adequate medical attention. Figure 10.8 shows the schematic of PM and its toxic effects on the human respiratory system. The explanation of various consequent stages is given in the following sections (Kim et al. 2017).

10.4.1 *Oxidative Stress*

The imbalance in the body occurs due to the oxidation and reduction processes, which is referred to as 'oxidative stress' (Gomes and Florida-James 2014). Due to

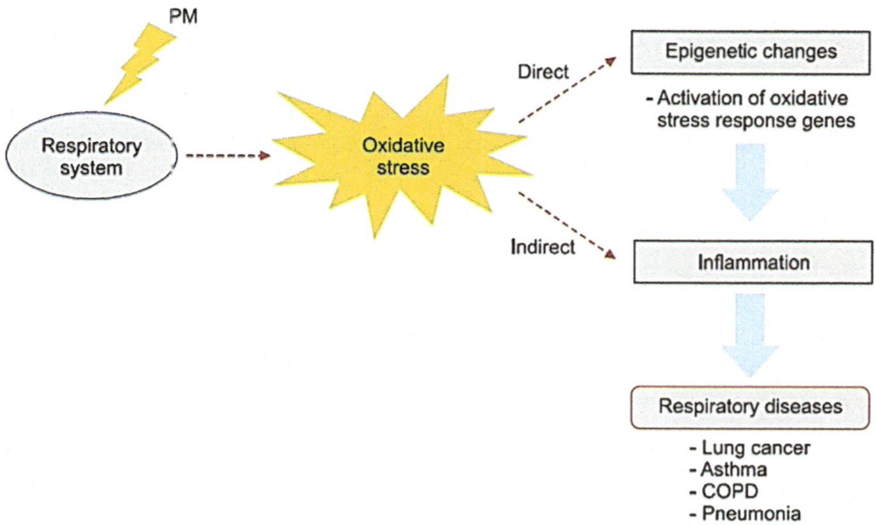

Fig. 10.8 PM and its harmful effects on the respiratory system (Kim et al. 2017)

the oxidation phenomena, reactive oxygen species (ROS) react with other molecules. The antioxidant mechanism of the body, which acts as a defense mechanism against ROS, is generally not able to overcome the total effect of ROS. Oxidative stress occurs primarily due to a high rate of ROS. The current situation can lead to catastrophic impacts on the body such as malfunctioning of cells or even the death of the cells. There are various diseases such as cancers of different types, neurological disorders, heart diseases, and pulmonary diseases (chronic obstructive pulmonary disease: COPD), which occur due to oxidative stress (Powers et al. 2004).

10.4.2 Inflammation

The term inflammation is a reaction response of the human body that comes into action when the immune cells of the body observe a pattern recognition receptor. It is a mechanism to protect the human body, which deals with immune cells (Kim et al. 2017; Lee et al. 2004). Different studies concluded that the inflammatory response is connected to the amount of PM inhaled (Jeon et al. 2011). Studies on inflammation (Dong et al. 1996; Long et al. 2001) reported that the controller, which is a bacteria-derived endotoxin bound to PM is responsible for lung injuries. When Fe and Cu components are present in the PM, they tend to induce oxidative stress and are responsible for the development of inflammatory diseases like asthma and lung cancer (Harrison and Yin 2000).

10.4.3 Epigenetic Change

An epigenetic change defines a molecular level variation that affects the gene style, without any change in the sequence of primary DNA (Loke et al. 2015). An epigenetic change is connected to a biological change in the body; hence, it affects the gene expression. Generally, environmental factors such as aging and diet are the primary causes of epigenetic change (Fraga et al. 2005). Recent studies (Kamata et al. 2005; Carmona et al. 2014) on human health related to PM confirmed that along with oxidative stress and redox status, PM also regulates epigenetic changes (Kim et al. 2017).

Therefore, oxidative stress, inflammation, and epigenetic changes in human health play an essential role in the context of respiratory diseases. Table 10.1 lists the respiratory illnesses induced by particulates, as reported by different researchers.

Table 10.1 PM and its connected respiratory diseases (Kim et al. 2017)

Author/s	Description	Related process	Related disease	Gene name
Hemmingsen et al. (2015)	8-oxoguanine DNA glycosylase	Oxidative stress, inflammation	Lung cancer, asthma, COPD	OGG1
Sun et al. (2012)	Toll-like receptor 4	Oxidative stress, inflammation	Lung cancer, asthma, COPD	TLR4
Bellavia et al. (2013)	Cytochrome P450 family 1 subfamily, a member 1	Oxidative stress, inflammation	Lung cancer, asthma	CYP1A1
	Interferon gamma	Oxidative stress, inflammation	Lung cancer, asthma, COPD	IFNG
Chu et al. (2016)	Tumor necrosis factor	Oxidative stress, inflammation	Lung cancer, asthma, COPD	TNF
	Colony stimulating factor 2	Oxidative stress, inflammation	Lung cancer, asthma, COPD	CSF2
	C-X-C motif chemokine ligand 8	Oxidative stress, inflammation	Lung cancer, asthma, COPD	CXCL8
Wang et al. (2012)	Heme oxygenase 1	Oxidative stress, inflammation	Lung cancer, asthma, COPD	HMOX1
	Solute carrier family 7 member 11	Oxidative stress, inflammation	Asthma	SLC7A11
	Stanniocalcin 2	Oxidative stress	Lung cancer	STC2
	Fission, mitochondrial 1	Oxidative stress	COPD	FIS1

10.5 Particulate Emissions from Vehicles

PM is one of the major pollutants among several others responsible for air quality deterioration. PM exhibits the most significant impact on the human health. Vehicular emissions are the biggest source of PM, especially microparticles (Onursal and Gautam 1997). Over the last several years, an increase in the number of stationary and non-stationary engines has increased and hence the concentration of PM. Exposure to such a high level of environmental particulates is responsible for various health problems and premature deaths (Lanki et al. 2006). Diesel engines generally emit nano-size PM. Hence, these PM have a higher tendency to penetrate the respiratory system. Warnatz et al. (2006) discussed the effects of exposure to fine PM on the human health. They reported that PM are emitted in relatively much higher numbers from the diesel engine compared to their gasoline counterparts. These particles get deposited in the alveolar region of the lungs since most of them have a diameter ranging from 20–200 nm.

In another study conducted by Tsurutani et al. (1995), they reported that long-term exposure to diesel emissions could increase the probability of lung cancer and bladder cancer by 20–50%. Many other studies (Environmental Protection Agency, U.S. 2002; Integrated Risk Information System (IRIS) 2003) conducted in recent years also reported that exposure to diesel emissions could lead to various other health issues such as acute eye infection, nausea, phlegm, cough. The soluble part of PM can lead to problems such as a tumor in the lungs or some stress in the lung region (Mauderly 1992). It is widely reported that PM can also contain trace metals such as Ca, Fe, Mg. The deposition of Fe traces in the lungs could produce free radicals, which are responsible for acute and chronic lung injuries (Valavanidis et al. 2000). Different studies reported that particulates emitted by the diesel engine might lead to allergic problems and finally some critical illness related to the respiratory system (Schwartz 2001). Also, long-term exposure to diesel particulates may lead to various health issues such as respiratory illness and asthma among the children (Lapuerta et al. 2008).

Chaichan (2015) conducted experiments to assess the effect of engine load and hydrogen on the emission of PM at 1500 rpm engine speed. Progressive addition of hydrogen in the engine reduced the PM emissions (Fig. 10.9). Initially, the PM level decreased with increasing engine load, but between 20 and 60 kN/m^2 the PM concentration became stable. Also, beyond 60 kN/m^2 engine load, the PM level increased rapidly.

Yusop et al. (2014) investigated the effect of palm methyl ester (PME) blended with methanol and biodiesel on the concentration of PM. The reduction in the PM concentration was higher when a blend of 10% methanol (PME20M10) fueled engine compared to the engine using other test fuels at a specific engine load (Fig. 10.10).

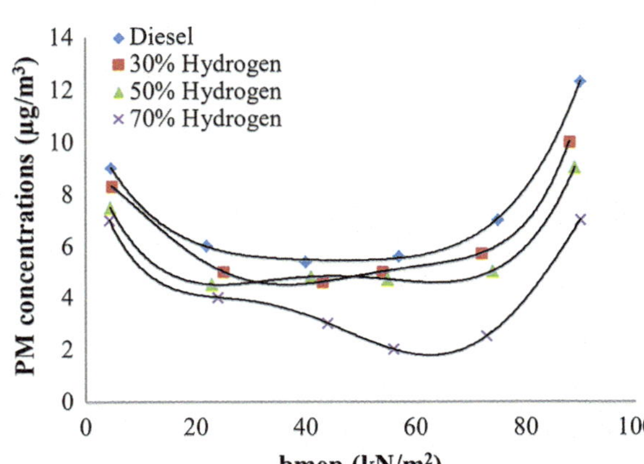

Fig. 10.9 Effect of hydrogen on PM emissions (Chaichan 2015)

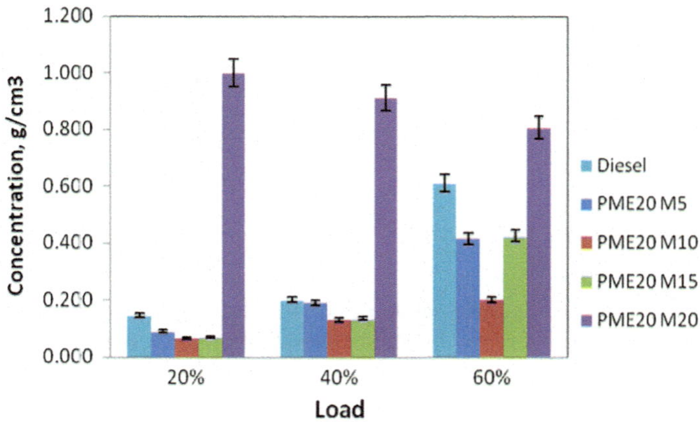

Fig. 10.10 Effect of methanol addition on PM emissions (Yusop et al. 2014)

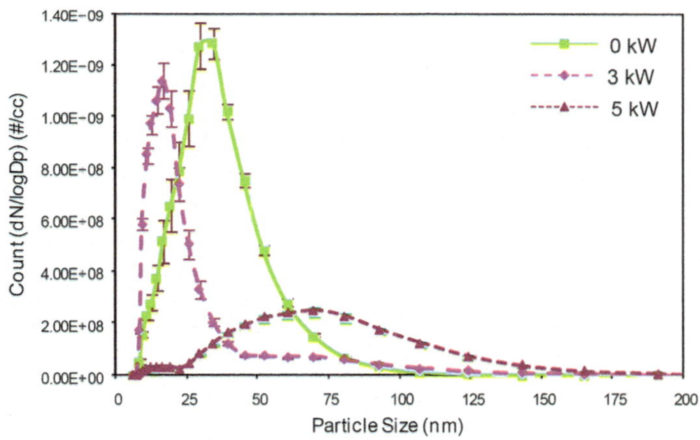

Fig. 10.11 Effect of engine load on number and size of particulate emissions (Srivastava et al. 2011)

In another study, Srivastava et al. (2011) investigated the influence of engine load on the size and number of PM emissions at constant engine speed. Figure 10.11 shows that with increasing engine load, the particulate number emissions decreased.

There are several other studies, which explore the use of biodiesel and its effect on the PM emissions. The PM emissions generally reduce with biodiesel usage in an engine. However, they become ultrafine in size, which may possibly be more dangerous. The impact of these ultrafine PM may be more severe on the human health and on the environment because the smaller particles have the capability to penetrate deeper into the respiratory system (Lapuerta et al. 2008). There are several hazardous issues reported in the literature due to exposure to diesel PM. Therefore, it is necessary to develop methods and techniques to effectively control the emission of diesel PM as the earliest in order to protect both humans and the environment.

10.6 Conclusions

PM emissions from IC engines and direct discharge into the environment are hazardous. Many researchers have demonstrated that prolonged exposure to exhaust emissions cause critical health issues such as cancer, respiratory problems, asthma. The chemical composition of fuel influences the size and amount of PM produced by the IC engines. The fuels with a low carbon-to-weight ratio as well as low aromatic content can reduce the PM emissions substantially. There are methods and techniques (pre- and post-combustion), which are used for reducing the concentration of PM emitted by modern IC engines. Pre-combustion PM-control strategies involve changing the fuel injection parameters, such as fuel injection pressure in diesel engines, injection timings, use of multiple injections, which reduce the PM emissions to a great extent. Similarly, use of common rail direct injection (CRDi) technology reduces both PM and oxides of nitrogen emissions from the engine. Post-combustion technology to control PM emissions includes diesel particulate filter (DPF), and diesel oxidation catalysts (DOC), which also help in compliances with existing emissions standards. To deal with exhaust emissions and compliance with the emission norms, exhaust gas after-treatment is preferred compared to pre-combustion methods.

References

Aalapati S, Ganapathy S, Manapuram S, Anumolu G, Prakya BM (2014) Toxicity and bio-accumulation of inhaled cerium oxide nanoparticles in CD1 mice. Nanotoxicology 8 (7):786–798

http://www.thecmmgroup.com/know-thy-enemy-particulate-matter/. Accessed on 09 July 2018

Bellavia A, Urch B, Speck M, Brook RD, Scott JA, Albetti B, Behbod B, North M, Valeri L, Bertazzi PA, Silverman F (2013) DNA hypomethylation, ambient particulate matter, and increased blood pressure: findings from controlled human exposure experiments. J Am Heart Assoc 2(3):e000212

Bruce CW, Stromberg TF, Gurton KP, Mozer JB (1991) Trans-spectral absorption and scattering of electromagnetic radiation by diesel soot. Appl Opt 30(12):1537–1546

Carmona JJ, Sofer T, Hutchinson J, Cantone L, Coull B, Maity A, Vokonas P, Lin X, Schwartz J, Baccarelli AA (2014) Short-term airborne particulate matter exposure alters the epigenetic landscape of human genes associated with the mitogen-activated protein kinase network: a cross-sectional study. Environ Health 13(1):94

Chaichan MT (2015) The effects of hydrogen addition to diesel fuel on the emitted particulate matters. Int J Sci Eng Res 6(6):1081–1087

Chu JH, Hart JE, Chhabra D, Garshick E, Raby BA, Laden F (2016) Gene expression network analyses in response to air pollution exposures in the trucking industry. Environ Health 15 (1):101

D'Amato G, Cecchi L, D'amato M, Liccardi G (2010) Urban air pollution and climate change as environmental risk factors of respiratory allergy: an update. J Invest Allergol Clin Immunol 20 (2):95–102

Dong W, Lewtas J, Luster MI (1996) Role of endotoxin in tumor necrosis factor a expression from alveolar macrophages treated with urban air particles. Exp Lung Res 22(5):577–592

Environmental Protection Agency, U.S. (2002) Health assessment document for diesel engine exhaust. EPA/600/8-90/057 F

Forouzanfar MH, Afshin A, Alexander LT, Anderson HR, Bhutta ZA, Biryukov S, Brauer M, Burnett R, Cercy K, Charlson FJ, Cohen AJ (2016) Global, regional, and national comparative risk assessment of 79 behavioural, environmental and occupational, and metabolic risks or clusters of risks, 1990–2015: a systematic analysis for the Global Burden of Disease Study 2015. Lancet 388(10053):1659–1724

Fraga MF, Ballestar E, Paz MF, Ropero S, Setien F, Ballestar ML, Heine-Suñer D, Cigudosa JC, Urioste M, Benitez J, Boix-Chornet M (2005) Epigenetic differences arise during the lifetime of monozygotic twins. Proc Natl Acad Sci 102(30):10604–10609

Gomes EC, Florida-James G (2014) Lung inflammation, oxidative stress and air pollution. In: Lung Inflammation. InTech

https://www.conserve-energy-future.com/causes-and-effects-of-particulate-matter.php. Accessed on 09 July 2018

https://www.epa.gov/pm-pollution/particulate-matter-pm-basics. Accessed on 09 July 2018

Harrison RM, Yin J (2000) Particulate matter in the atmosphere: which particle properties are important for its effects on health? Sci Total Environ 249(1–3):85–101

Hauser R, Eisen EA, Pothier L, Christiani DC (2001) A prospective study of lung function among boilermaker construction workers exposed to combustion particulates. Am J Ind Med 39 (5):454–462

Haynes BS, Wagner HG (1981) Soot formation. Prog Energy Combust Sci 7(4):229–273

Haynes BS, Bartok W, Sarofim AF (eds) Fossil fuel combustion. Wiley, New York, pp 261–326

Health Effects Institute (2018). State of global air 2018. Special Report. Health Effects Institute, Boston

Hemmingsen JG, Jantzen K, Møller P, Loft S (2015) No oxidative stress or DNA damage in peripheral blood mononuclear cells after exposure to particles from urban street air in overweight elderly. Mutagenesis 30(5):635–642

Integrated Risk Information System (IRIS) (2003) Diesel engine exhaust. Retrieved from www.epa.gov/iris/subst/0642.htm

Jang AS (2012) Particulate air pollutants and respiratory diseases. In: Air pollution—a comprehensive perspective. InTech

Jeon YM, Son BS, Lee MY (2011) Proteomic identification of the differentially expressed proteins in human lung epithelial cells by airborne particulate matter. J Appl Toxicol 31(1):45–52

Kamata H, Honda SI, Maeda S, Chang L, Hirata H, Karin M (2005) Reactive oxygen species promote TNFα-induced death and sustained JNK activation by inhibiting MAP kinase phosphatases. Cell 120(5):649–661

Kim HJ, Choi MG, Park MK, Seo YR (2017) Predictive and prognostic biomarkers of respiratory diseases due to particulate matter exposure. J Cancer Prev 22(1):6

Lanki T, Pekkanen J, Aalto P, Elosua R, Berglind N, D'Ippoliti D, Kulmala M, Nyberg F, Peters A, Picciotto S, Salomaa V (2006) Associations of traffic related air pollutants with hospitalisation for first acute myocardial infarction: the HEAPSS study. Occup Environ Med 63(12):844–851

Lapuerta M, Armas O, Rodriguez-Fernandez J (2008) Effect of biodiesel fuels on diesel engine emissions. Prog Energy Combust Sci 34(2):198–223

Lee KO, Cole R, Sekar R, Choi MY, Zhu J, Kang J, Bae C (2001) Detailed characterization of morphology and dimensions of diesel particulates via thermophoretic sampling (No. 2001-01-3572). SAE technical paper

Lee JY, Zhao L, Youn HS, Weatherill AR, Tapping R, Feng L, Lee WH, Fitzgerald KA, Hwang DH (2004) Saturated fatty acid activates but polyunsaturated fatty acid inhibits Toll-like receptor 2 dimerized with Toll-like receptor 6 or 1. J Biol Chem 279(17):16971–16979

Loke YJ, Hannan AJ, Craig JM (2015) The role of epigenetic change in autism spectrum disorders. Front Neurol 6:107

Long CM, Suh HH, Kobzik L, Catalano PJ, Ning YY, Koutrakis P (2001) A pilot investigation of the relative toxicity of indoor and outdoor fine particles: in vitro effects of endotoxin and other particulate properties. Environ Health Perspect 109(10):1019

Mauderly JL (1992) Diesel exhaust. In: Lippmamr M (ed) Environmental toxicants. Human exposures and their health effects. Van Nostrand-Reinhold, New York, pp 119–62 (1992)

Mohankumar S, Senthilkumar P (2017) Particulate matter formation and its control methodologies for diesel engine: a comprehensive review. Renew Sustain Energy Rev 80:1227–1238

National Research Council (2004) Research priorities for airborne particulate matter: IV. In: Continuing research progress, vol 4. National Academies Press

Onursal B, Gautam SP (1997) Vehicular air pollution: experiences from seven Latin American urban centers. The World Bank

Pinson JA, Ni T, Litzinger TA (1994) Quantitative imaging study of the effects of intake air temperature on soot evolution in an optically-accessible DI diesel engine. SAE Trans 1773–1788

Powers SK, Deruisseau KC, Quindry J, Hamilton KL (2004) Dietary antioxidants and exercise. J Sports Sci 22(1):81–94

Pruss-Ustun A, Corvalán C (2006) Preventing disease through healthy environments: towards an estimate of the environmental burden of disease. World Health Organization

Schwartz DA (2001) Does inhalation of endotoxin cause asthma? Am J Respir Crit Care Med 163 (2):305–306

Smith OI (1981) Fundamentals of soot formation in flames with application to diesel engine particulate emissions. Prog Energy Combust Sci 7(4):275–291

Srivastava DK, Agarwal AK, Gupta T (2011) Effect of engine load on size and number distribution of particulate matter emitted from a direct injection compression ignition engine. Aerosol Air Qual Res 11(7):915–920

Sun H, Shamy M, Kluz T, Muñoz AB, Zhong M, Laulicht F, Alghamdi MA, Khoder MI, Chen LC, Costa M (2012) Gene expression profiling and pathway analysis of human bronchial epithelial cells exposed to airborne particulate matter collected from Saudi Arabia. Toxicol Appl Pharmacol 265(2):147–157

Tree DR, Foster DE (1994) Optical measurements of soot particle size, number density, and temperature in a direct injection diesel engine as a function of speed and load. SAE Trans 318–330 (1994)

Tsurutani K, Takei Y, Fujimoto Y, Matsudaira J, Kumamoto M (1995) The effects of fuel properties and oxygenates on diesel exhaust emissions (No. 952349). SAE technical paper

Valavanidis A, Salika A, Theodoropoulou A (2000) Generation of hydroxyl radicals by urban suspended particulate air matter. The role of iron ions. Atmos Environ 34(15):2379–2386

Wang T, GN Garcia J, Zhang W (2012) Epigenetic regulation in particulate matter-mediated cardiopulmonary toxicities: a systems biology perspective. Cur Pharmacogenomics Pers Med (Formerly Curr Pharmacogenomics) 10(4):314–321

Warnatz J, Maas U, Dibble RW (2006) Combustion: physical and chemical fundamentals, modeling and simulation, experiments, pollutant formation. Springer, Berlin

WHO (2003) Health aspects of air pollution with particulate matter, ozone and nitrogen dioxide. World Health Organization (WHO), Regional Office for Europe, Copenhagen
Williams PT, Andrews GE, Bartle KD (1987) The role of lubricating oil in diesel particulate and particulate PAH emissions. SAE Trans 457–465
Yusop AF, Mamat R, Mat Yasin MH, Ali OM (2014) Effects of particulate matter emissions of diesel engine using diesel–methanol blends. J Mech Eng Sci 6:959–967

Printed by Printforce, the Netherlands